Security Systems and Intruder Alarms

Security Systems and Intruder Alarms

Second edition

Vivian Capel

Newnes

OXFORD AUCKLAND BOSTON JOHANNESBURG MELBOURNE NEW DELHI

Newnes
An imprint of Butterworth-Heinemann
Linacre House, Jordan Hill, Oxford OX2 8DP
225 Wildwood Avenue, Woburn, MA 01801–2041
A division of Reed Educational and Professional Publishing Ltd

A member of the Reed Elsevier plc group

First published 1989
Reprinted 1992, 1993, 1994, 1995, 1997
Second edition 1999
Transferred to Digital Printing 2004

British Library Cataloguing in Publication Data
A catalogue record for this book is available from the British Library

ISBN 0 7506 4236 X

Composition by Genesis Typesetting, Laser Quay, Rochester, Kent

Contents

Preface

It is a sad comment on our present civilization that one of the main boom 'industries' is that of crime. In a typical year half a million commercial premises are broken into with losses exceeding £70 million.

All businesses are vulnerable, small and large, especially those that carry expensive stock such as jewellers, furriers and wine and spirit merchants. To these can now be added the chemist, as drug addiction has given his stock special value.

Even if little is taken, much damage can be caused, not only by breaking in, but by acts of vandalism which are common when the haul is poor. The destruction of files and records could be a major loss to most businesses.

No less worrying is the prevalence of shop-lifting. Over a quarter of a million offences are recorded each year with a loss of some £10 million, but this is believed to be just the tip of the iceberg, as by far the majority goes undetected. Surveys suggest that up to one in every fifty customers steals something, but the detection rate is as low as from one in a hundred to one in a thousand. This could put losses at a staggering £1 billion.

Even more depressing is the incidence of employee theft because employees are trusted and are in a position to get away with far more. Some 20,000 cases a year are recorded, but the average amount stolen is over £900. Total recorded losses are thus £18 million, nearly twice as much as that listed for shop-lifting. But again, the evidence is that most is undetected and the actual sums lost are far greater.

The director general of the CBI once put the total annual cost of crime to British industry at a colossal £5 billion, more than spent on all non-military research and development.

Another major danger is fire which destroys property valued at £450 million each year. Insurance can never compensate for the upheaval, unusable premises, loss of business and destruction of records that result from a major fire, to say nothing of the danger to life.

Added to these are the possibilities of computer and company fraud, public liability claims and the risk of fire and arson. A veritable minefield of losses, hazards and risks. To help combat these, this book has been written. Though concentrating on alarm systems, each of the other problems are also carefully considered.

In the interests of clarity the book is divided into two parts. First comes the General Section which aims to dispel the mystery surrounding intruder and fire alarms, show how they work, what they do and how systems can be planned or a proposed plan checked. It goes on to describe countermeasures against the other hazards thereby offering all-round defence against criminal and other loss. The limited technical content is fully explained.

Most of the technical information though, has been segregated into the final Technical Section which is intended for installation engineers and those who want to dig deeper into the technical details. It covers installing, testing, fault-finding and maintenance of intruder and fire alarm systems as well as reference data, and summaries of the National Supervisory Council for Intruder Alarms code and the relevant British Standards.

Since the last edition, new detection devices and other items have been introduced to help the business owner in the fight against crime. These have been included in the text in the appropriate sections, as well as updates on existing ones.

Regrettably, many new scams and dangers have also reared their heads. Among these are: ghost employees, petty cash fiddles, false CVs and references, fraud, false invoicing, phone hackers, theft of whole companies (yes, unbelievably, this is happening), new legal liabilities, and violence against staff. These are also discussed and ways suggested to combat them in three new chapters.

A further hazard is security systems that have been fitted by 'cowboy' installers, and fall short of providing the security desired. A chapter describes an actual system, and invites the reader to test his knowledge by identifying its weak points. The answers are given at the end of the chapter.

It is sincerely hoped that this book will make a significant contribution to the reduction of crime-related losses in business.

Vivian Capel

Acknowledgements

The author would like to thank the following firms for their invaluable help in supplying product information for this book:

Al Security, Cambridge.
ATH Ltd, Bristol.
C-Tec Security, Wigan.
RS Components Ltd, Corby.
TP Security, Bristol.

Part One

General Section

1 Alarm system requirements

Our first step in considering an alarm system is to define precisely what is required of it. The principal function is usually considered to be to warn or inform others that an intrusion has taken place in the premises concerned. This really is of secondary importance, especially as is often the case, little notice is taken of it due to the prevalence of false alarms.

Rather, the prime function is to deter. Often the very sight of an alarm bell outside the building and other signs of electronic detection is sufficient to warn off the would-be intruder. Even though alarm bells or sirens are often ignored, they are bound to attract attention, and the last thing an intruder wants is publicity. His aim is to get in and out unobserved and unrecognized. If there is any chance of his being detected, he will look elsewhere for a less risky subject for his attentions.

In most cases, entry is attempted at the rear of the premises where security is often poor and there is less chance of being observed. If the alarm system is set off, it takes a stronger nerve than most intruders have, to enter with the alarm sounding. The immediate reaction is that of panic and an overwhelming desire to get away quickly. Usually the louder and more strident the noise, the greater the panic it produces.

This of course is just what is desired. Some control units allow the audible alarm sounding to be delayed while immediately signalling an intrusion to the local police station. The object is to catch the culprit red-handed, but the wisdom of this is very dubious. It is far better to scare off the intruder, than to allow him to enter in the hope that he may be caught. He may still escape before the police arrive, and could do considerable damage in the meantime. A better rule is: *Prevention is better than Apprehension.*

To accomplish this effectively, the alarm should be loud and clearly heard wherever an entrance is possible. This means that a sounding device should be provided at the rear of the premises as well as at the front. It is not often done, but is well worth the small extra cost.

Reliability

A vital factor is reliability. If the system is out of action for even a short time, that could be the very time the intruder strikes. Reliability is dependent on first, the control unit, its design, construction and quality of components, and second, on the installation, its wiring and ancillary equipment including the sounding device.

The reliability of the control unit cannot be assessed by a potential user. It would require experience of a large number of the same model to determine, something that only the large installation companies would have. If there is a poor record of reliability for any particular model it is unlikely they would continue to stock or supply it. A model that has been on the market for some while and is still readily available from several firms is thus likely to be a better risk than a new one, however attractive its features may be.

Actually, all electronic components are liable to failure, and a failure rate per thousand samples plus a mean time to expected failure is assessed by exhaustive testing and is specified by component manufacturers. There is a British Standard as well as IEC standards that define methods of expressing failure rates.

This shows that 100 per cent reliability for components is unattainable and any claims for such should be treated with the utmost scepticism. It follows from this that the more components the control unit has and the more complex it is, the more likely, statistically, it is to fail. Complex units having many features are now offered at quite reasonable prices compared to what they were at one time. This is partly due to the use of dedicated (specially designed) silicon chips which carry out most of the functions. However, a host of features you will never use could be obtained at the cost of higher liability to failure.

This is not to say that the present generation of alarm units are basically unreliable, only that the chances of a breakdown are greater with a more complex unit. Having made this point it should be said that the most likely cause of a fault is in the installation, in the wiring, sensors, or in the use of unsuitable sensors. This is where a reputable installation firm scores, they have to, or should, abide by the National Supervisory Council for Intruder Alarms (NSCIA) code of practice which ensures the highest standards of installation. However, there is no reason why a competent DIY man should not design a system and install it to the same high standards. Further chapters in this book show how this can be done as well as describing the NSCIA and British standards governing such installations.

False alarms

One of the biggest problems with alarm systems is that of the false alarm. In the Metropolitan police area no less than 98 per cent of the call-outs are for false alarms. Understandably the police see this as a considerable waste of their time and resources and some forces have been forced to lay down conditions as to how far they will respond. More than five false alarms in any month and the owner is warned; after three months of nuisance calls there will be no further police response to an alarm.

Apart from police involvement, the goodwill of neighbours will inevitably be jeopardized and strained by frequent false alarms, to say nothing of the trauma experienced by the keyholder who is frequently hauled from his bed to answer calls in the middle of the night. It is the high incidence of false alarms that have resulted in the general ignoring of a sounding alarm.

A trick that has been used by some burglars is to try to deliberately set off the alarm in some way without making an obvious entry or disguising it, then wait nearby. The keyholder arrives, makes a quick check of the premises and concludes that it is another false alarm. Assuming a fault on the system and not wishing to be summoned again that night or to antagonize the police, he switches the system off. The NSCIA code of practice actually recommends not switching the system on again after a false alarm. So, when the fuss has died down the intruder then makes his entry, secure in the knowledge that he will not be disturbed.

This ruse can only work if the system is prone to false alarms and can be triggered from the outside or without causing obvious damage. Illustrating how over-sensitive some systems are, is the fact that one night in October 1987, a security firm in the London area logged 3000 calls in a period when they would normally receive 8 to 12 calls. The police computer handling incoming calls broke down under the strain. The reason? that was the night of the hurricane which swept the country and the alarms were due entirely to the high winds and their effects.

It can be seen then, that false alarms must be avoided above all else, but in a way that does not compromise security. It can be done, as there are a large number of effective installations that rarely if ever experience one. Human error is sometimes to blame and little can be done about that except to stress extreme care on everyone concerned. Most of the trouble lies in the installation of the system or the use of unsuitable sensors. This aspect will therefore be stressed in our subsequent considerations.

Cost effectiveness

One factor that is sometimes overlooked is the balance between the level of appropriate security and the cost. It is possible to fortify premises to a standard that would do justice to a bank, at a cost of tens of thousands of pounds to install and a considerable recurring sum to check and maintain. This may indeed be desirable for a large jeweller's premises or a warehouse of valuable items, but would it be worthwhile for a corner grocer's shop?

Security costs money, so buying more security than you really need is not good business practice, although it is always better to err on the safe side. A really determined professional crook will gain access almost anywhere and go to infinite pains to do so. He can defeat even the best security systems. However, such individuals are fortunately only a small percentage of the total number of intruders. Most are opportunist thieves looking for an easy 'job', many being local ne'er-do-wells, who are often youths or even children.

Quite a high degree of security can be achieved, sufficient to defeat the efforts of such individuals, at moderate cost. However, the mistake must be avoided of cutting costs by making part of the premises secure while neglecting other parts. The old adage about the strength of a chain being that of its weakest link is very true here. A weak point in the system, whether in securing premises, deterring shop lifting or eliminating staff pilfering will quickly be discovered and exploited. Money spent on other areas of security will then be like stopping just one hole in a bucket full of holes.

In the case of security installations, a question to be considered is whether to call in a professional firm of installers or to attempt a DIY job. Large premises, such as factories or warehouses, are nearly always best tackled by the professionals. Even with these though, a good working knowledge of security systems is worthwhile so that different systems can be assessed.

If the security system can be DIY installed, the costs will be considerably less, but it must be done properly otherwise it may engender a false sense of security, and be prone to annoying false alarms. However, only approved professional installations usually qualify for insurance premium reductions.

Having established the essential system requirements, we can now go on to take a look at the basic alarm circuit and its elements.

2 The basic alarm system

Although an alarm installation in large commercial premises may be a complex affair it can be broken down into four basic essentials. First, there is the sounding device which in most cases is a bell, but frequently is a siren. Second, there is a power supply which can be derived from the mains with a battery back-up, or in smaller systems can be just a battery. Next, there are switches which are activated by the intruder. These can take many forms and so are more usually described as sensors. Finally there is a master control which switches the whole circuit on or off, or selects test and other modes of operation.

The basic circuit

In Figure 1 we have the simplest possible alarm system containing these four elements. It is similar to an ordinary door bell circuit except that several switches are connected in parallel across each other. There is also a master switch. If any one of the sensor switches is closed the alarm bell will ring. There is no limit to the number of sensors that can be wired into the circuit, and they can be of different types, such as door switches that operate when a door is opened, or pressure pads under carpeting that close when trodden on.

As it stands, there is a very serious limitation to this circuit. If the bell rings when a door is opened, it can be stopped simply by closing the door again. Or, if the alarm is activated by stepping on the carpet, it will stop as

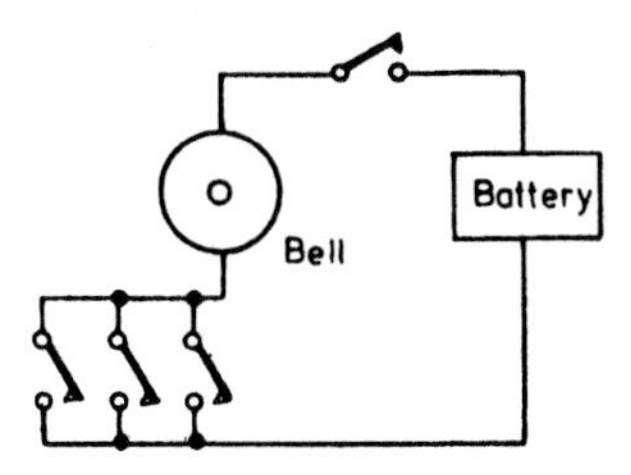

Figure 1 Basic alarm circuit consisting of bell, battery, control switch and parallel sensors.

soon as the pressure is released by stepping off. Obviously such an arrangement is of little use as a security device.

What is needed is a means whereby the alarm, which when started, latches on and continues to sound irrespective of what subsequently happens to the sensor switches. Only throwing the master switch can then silence it. A latching arrangement is therefore an essential part of all intruder alarm systems.

The principle can be illustrated by a simple mechanical arrangement that was used in the pre-electronic era and is shown in Figure 2. It consisted of a metal rod that rested on the bell striker against the force of a spring. When the bell was activated, the striker moved inward so releasing the rod which was pulled by the spring into contact with a metal cradle beneath.

Figure 2 Simple mechanical latching system. The metal rod rests on the bell striker and is pulled down to engage with metal the contact by a spring when the striker moves. The rod and contact are connected across the sensor circuit and so keep the bell circuit closed.

The rod and cradle were connected across the sensor switches so that the circuit was completed even if all the sensors were then put in the off position. It was reset by means of a lever which lifted it clear of the cradle to rest again on the striker arm.

Before the advent of semiconductors, the method which was universally used and still forms a part of some systems, is the latching relay. This is a switch that is magnetically operated when an electric current is passed through a coil of wire. When the current ceases, the switch contacts spring back to their former position.

A simple arrangement is shown in Figure 3. The coil is connected across the bell and the relay switch across the sensors. When a sensor is closed, current flows through the bell and also the relay coil. The relay switch is thereby operated, so closing the circuit and maintaining it in the 'on' state. When the master switch is opened, the relay current ceases and its switch resets.

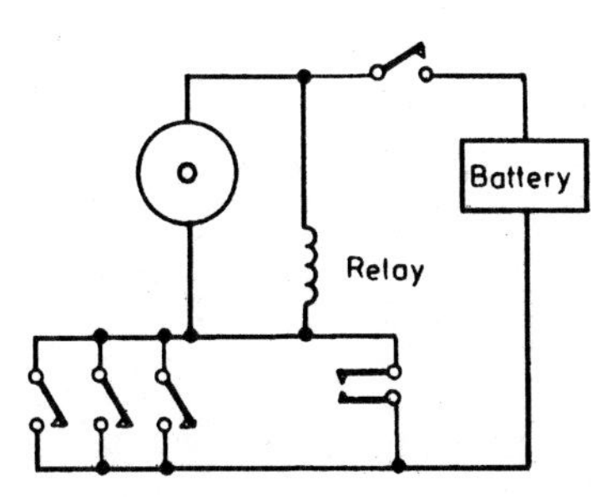

Figure 3 Using a relay to latch the circuit on. When a sensor closes, current flows through the bell and relay coil. The relay switch closes, shorting the sensor circuit, so maintaining the alarm even if the original sensor is then opened.

Closed loops

The circuit shown in Figure 3 is what is known as an 'open' circuit, that is, all the sensor switches are normally in the open position, and when activated they close. This arrangement has certain drawbacks. It is vulnerable to tampering, which is especially important with commercial or business premises where there is common access during business hours. A wire to one of the sensors could be cut and there would be no means of detecting it.

The other snag is that testing can only be done by actually operating each switch in turn, which would be very inconvenient if the alarm was sounded each time. Even if a test circuit was devised to prevent this, it would still be a laborious and time consuming task to check each sensor individually.

A system test would thus be unlikely to be carried out very often, and faults both deliberate and the result of accidental damage, could pass unnoticed for long periods with consequent loss of protection.

The alternative which avoids these difficulties is the closed loop, a basic circuit of which is shown in Figure 4. Instead of the sensors being normally open, they are normally closed and are connected in a series loop as shown, rather than being wired in parallel.

If any of the sensors are actuated, the contacts open, thus open-circuiting the entire loop. The control unit continually circulates a small current around the loop when in the on-guard condition. Any break in the loop stops the current which is interpreted as an alarm situation and so triggers the sounding circuit.

In the case of the circuit in Figure 4, the relay is continuously energized by the current passing through the loop. The associated switch is known as a single-pole changeover, it has three contacts, of which one is common. This contacts each of the others in turn as the switch is actuated.

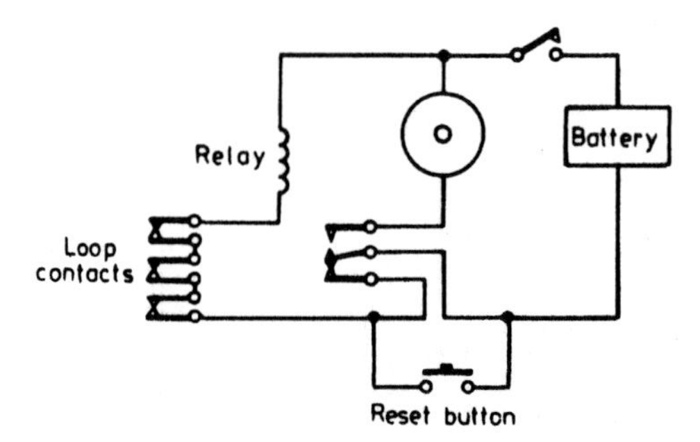

Figure 4 Relay latching circuit with closed loop. Coil is continuously energized holding the switch open. If a sensor opens, the relay is de-energized and the switch closes to sound the bell.

When energized, one pair of contacts close, so passing current through the loop and relay coil. Any break in this circuit caused by a sensor switch opening, de-energizes the relay, so opening the first pair of contacts and closing the others. The second pair close the bell circuit thereby sounding it. As the first pair are now opened, any closure of the loop cannot re-energize the relay and stop the alarm. The circuit is thus latched in the alarm position.

To reset the circuit, the reset button is momentarily depressed. This is connected across the first pair of relay contacts and completes the loop circuit so energizing the relay. Once it is energized, the first contacts are held in the closed position thus continuing the loop and the energizing current.

The snag with this circuit is that the relay current is flowing all the time the circuit is on guard or 'armed' in the terminology of intruder alarms. Relays that were originally used, took a heavy current, and even large capacity batteries had a short life. More recent relay types were far more frugal in their current demands, but still needed a fair amount that could drain a battery after a few days of continuous operation.

This may seem unimportant when the supply is obtained from the mains, but it is not when battery backup life is considered. If a mains fault tripped a circuit breaker in the building at the start of a long holiday period when the premises are unoccupied for a week or more, how long will the standby battery last? This is obviously an important factor.

The simple relay circuits originally used, disappeared when transistors came on the scene. Transistors were used to control the relay, as only a small current was needed to turn the transistor off, which when removed turned it on, thereby activating the relay. Now, the relay has been dispensed with altogether, its place being taken by semiconductor devices.

Advantages of the closed loop

If the loop has been cut or otherwise damaged and is open circuit, the alarm will sound or some indication will be given as soon as the system is switched on. Conversely, if it does *not* sound, then, providing there is no fault in the control box or sounder this indicates that the loop is intact. It thus has an inherent fail safe factor and is in effect tested each time the system is switched on. This contrasts with the parallel switch arrangement in which each sensor must be operated for testing.

System testing

It follows from this that the control box and the sounder can be tested by simply open-circuiting the loop at the control box. So if the alarm does not sound when it is switched on we know that the loop is in order, and if it then sounds when the loop is broken by a test switch on the control box, the rest of the system is working.

It is not necessary for the alarm to be actually sounded for test purposes. Most systems have a light that is switched on in place of the bell for testing and thus the whole system except the bell and its wiring can be silently tested each time the system is switched on. It is possible to test the bell and its wiring silently, by passing a small current through it, that is sufficient to operate an indicator circuit but not large enough to activate the bell. One system, once used this method, so making a total test with a single operation, but it is no longer available.

Some method of testing at least part of the system each time it is switched on, is essential if maximum security is to be maintained. Without it, a fault could develop or it could be deliberately put out of action leaving the premises unprotected until the trouble is discovered. Obviously, the more of the system that is tested, the greater the security.

Tamper-proofing

A simple two-wire loop is quite adequate for domestic alarm systems, where those having access to the premises when the alarm is off are members of the owner's family, or persons known to him. In the case of business premises open to casual visitors or to the public, tampering is a possibility. Ideally all parts of the system wiring should be buried or concealed, although this is not always practicable.

If the loop wiring has been cut, it becomes evident when the system is switched on as we have already seen. But this probably will be when the premises are being closed at night, a very inconvenient time to discover it.

It is far better to receive a warning when the actual damage is inflicted so that immediate steps can be taken to repair it, and possibly discover the culprit.

To protect the loop and give immediate warning of any tampering, a four-wire system is commonly used for business installations. One pair is the normal loop having all the switches connected in series with it, and the other is a loop which connects to blank terminals on each sensor, which are just used as connecting points, or as straight-through links.

Current is passed around the second protection loop continuously for 24 hours a day whether the system is switched on or off. As none passes through the actual sensor switches it is not affected by the normal comings and goings. If the wire is cut though, the loop is broken and an alarm is immediately sounded. This need not be the main system bells or sounders which could alarm genuine customers on the premises, but a smaller device in the supervisor's or security guard's office. The protection loop is usually non-latching because unlike the sensor loop there is no need to ensure it stays in the on condition, the damage will not repair itself!

The anti-tamper loop can be connected to any vulnerable part of the installation. The bell box can have a microswitch operated by a spring-loaded plunger positioned inside and held down by the cover. If the cover is removed, the switch is released and the protection loop open-circuited, so triggering the tamper alarm. A similar arrangement operates in the control box and some space protection sensors such as infra-red devices, ultrasonic and microwave detectors. An anti-tamper loop plus, detection loops, is thus the normal system circuit.

Where extra security is required in the case of valuable stock that could attract the attention of professional thieves, even the normal four-wire system incorporating a 24 hour protection loop may be considered insufficient. It is possible for tamperers to peel back the outer insulation and short out the detection loop wires with a pin thus disabling the associated sensor switch or switches. The detection loop pair would have to be identified, but shorting both pairs could be easily done without triggering the anti-tamper circuit. The problem would be to identify which were the pairs out of the four wires as there is no standard colour code.

Added protection could be afforded by means of sensors using single-pole changeover switches having three contacts, wherein a common contact makes with one contact while it breaks with the other. The pair that are normally closed are wired into a loop in the conventional manner and the odd one is connected to those of the other switches as in Figure 5. They are then taken to the open-circuit terminals of the control box. These are used for sensors, such as pressure mats, that are normally open-circuit and cannot be connected in a loop.

Should any of the normally-closed loop connections be shorted out by bridging the wiring, the sensor would still trigger the alarm, by activating

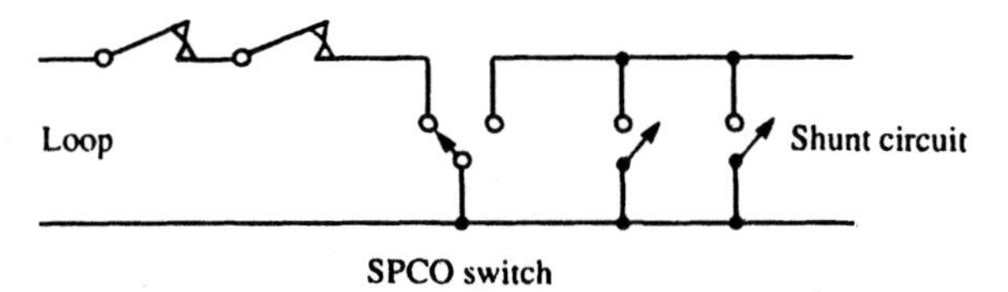

Figure 5 Single pole changeover switch connected either in normally-closed mode to a loop circuit, or in normally-open mode in a shunt circuit.

the open-circuit facility of the system. Furthermore the extra wiring serves to decrease the probability of a tamperer guessing the identity of the wires.

Changeover switch sensors are rarely used in spite of their added security, and are not easy to obtain, but there is an alternative.

Dual-purpose loop

The open-circuit facility to which pressure mats are connected requires separate wires back to the control box. So, if a mat and a loop sensor are situated some distance from the box, either a separate run must be made or extra cores included in the loop cable.

This can be obviated in systems that have a dual loop facility (Figure 6). A terminating resistor is included at the most distant part of the loop and higher value resistors are connected across each pair of series sensor contacts. Pressure mats are connected across the loop. A continuous current can circulate in the 'day' mode because the loop is completed by the sensor

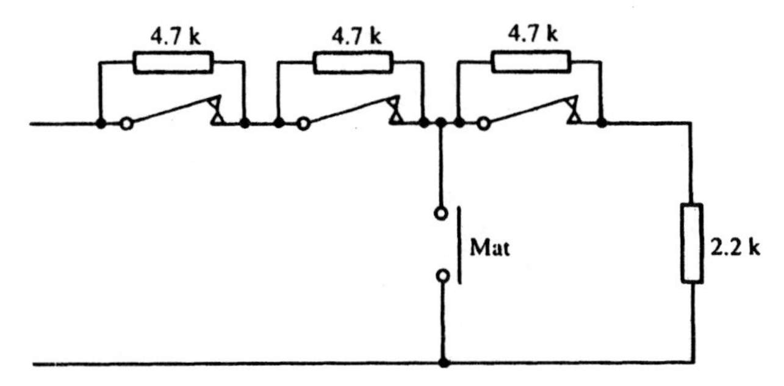

Figure 6 Dual-purpose loop with end-of-line resistor. Pressure mats can be connected across the circuit thereby causing a short-circuit when operated; series switches with shunt resistors can be wired in a loop, producing a high resistance when opened.

resistors irrespective of whether the contacts are open or closed. If the loop is broken, a tamper indication is triggered. Thus it is continually monitored without the need of a separate anti-tamper circuit.

In the 'night' mode, if a series sensor switch is operated the value becomes infinity because the loop is open-circuited, or if it is shorted out by a pressure mat being actuated, the value is zero. Either condition triggers the alarm. Thus, both normally open and normally closed sensors can be mixed.

Another common method is to wire the pressure mats from the detection loop to the protection circuit. With control units that are designed for this mode of connection, any short-circuit appearing between these loops initiates the alarm.

Zoning

It is sometimes necessary to protect part of a building that is unoccupied when another part is not, while at other times all parts are unoccupied and need protection. This may be the case when premises accommodate office, sales, and service departments that work different hours. Or, in the case of small businesses, there may be living accommodation in the same building as the business, such as a flat over a shop.

In such cases it is very useful to be able to switch on only part of the alarm circuit that covers the unoccupied areas. Even in many domestic situations this could be desirable, as intruders have entered the back of the house while the family were watching television at the front.

Systems can be divided into two or more zones, and the control units carry a specification as to the number of zones that can be served. Usually, one zone is activated all the time the system is switched on, while the others can be switched in as required. All sensors serving the first area to be unoccupied are connected to the first zone, while those protecting areas that are still occupied are linked to the other zones.

With small systems, particularly domestic ones, a form of zoning can be achieved with a single-zone control unit by simply installing a switch to short out part of the loop. It could be located near the control panel or at some other convenient point. To be secure, it would need to be hidden or disguised if installed remote from the panel, and itself be within the area still protected when it was switched off. It thus could not be operated by an intruder without activating the remaining sensors. A disadvantage is that it could be forgotten and left off when total protection was needed.

With large premises, zoning offers another major advantage. Most zoned control units have visible indicators to show which zone tripped the alarm. If more than one zone has been activated, many control panels indicate which was the first one. This enables a speedy identification of the area of actual or attempted entry. The indication remains even after the bell may have been silenced by a timer.

Zone identification is desirable, not only in the case of activation by an intruder, but also when there is a false alarm. The source of these is often difficult to determine, and a zone indication narrows the field considerably.

Panic button

All systems now have the facility of being triggered by a press button, even when they are switched off. This is generally known as a panic button and serves to summon assistance in an emergency, or scare off an attacker. The obvious application is at a cashier's desk or check-out point, but they can be installed anywhere where an attack is possible.

The buttons are usually of the normally-closed type that open when pressed and so can be connected to the 24 hour protection circuit. If this has been designed to sound a limited alarm to only the supervisor or security staff, then the panic button will do the same. Depending on the likely circumstances and possible time taken to summon help, it may be considered preferable to sound the full alarm and scare off the attacker. In this case a separate personal attack facility would be required at the control box to which the button should be wired. Many systems have this separate public address (PA) facility. Alternatively, the 24 hour protection loop could be put on full alarm status.

As some PA circuits are non-latching, the panic button itself needs to latch. It also needs to be defeat-proof so that the assailant cannot reset it and thus silence the alarm. This is done by making the button resettable only with a key. Most panic buttons are recessed to prevent accidental operation.

If several panic buttons are to be included in the system they can be connected in series in a loop. This confers the same advantages as the main sensor loop. All can be tested without individual operation by checking the loop continuity, and any break in the wiring is readily detected.

Exits and entrances

Having switched the system on, the problem now arises as to how to get out of the building. There are many possible methods and all of them have snags. The usual method is to have a special circuit called an exit loop to which sensors on the exit door, and any others that may be encountered on the way out are wired. The loop is subject to a time delay that can usually be set at the control box for any time from a few seconds up to several minutes.

Once the system has been switched on, the delay timer starts running. All sensors are immediately operational except those on the exit loop which remain inactive until the delay time has expired. After this any actuation of a sensor on the exit circuit starts another timer running, and when this has expired the alarm sounds unless the system has been switched off first.

This second timing period is to allow re-entry without tripping the alarm, and although in most cases the same time would be required to enter and switch off the system as would be needed to exit after switching on, the timers are generally (though not always) independently adjustable.

The occupier thus has a certain limited time both to exit and to enter and this could give rise to false alarms should he be unexpectedly delayed. For example, dropping keys or files which spilled their contents on the way out could cause the exit door to be open after the exit time had expired. This would not immediately sound the alarm but would start the entrance timer. Thus the alarm would sound some minutes (or however long the entrance time is) later, possibly when the occupier has passed out of hearing range.

To reduce this possibility, many systems have an internal buzzer which sounds when the exit timer is running but stops when the time has expired. So as long as the buzzer is sounding and can be heard from the exit door, it is safe to leave. Should it stop before the exit is made, the system must be switched off and switched on again. Should the buzzer stop a split second before the exit door is closed, the time interval may pass unnoticed and the entrance timer would start running. An apparently inexplicable false alarm would thereby result.

The buzzer usually also sounds during the entrance period. It then serves as a reminder that the system is still switched on and must be turned off immediately if an alarm is not to be sounded.

There is thus little to be gained in setting the exit and entrance timers to too-short a period. Every exit and entrance would then be a stressful event trying to beat the timer, and sooner or later an unexpected delay would occur. Intruders would very rarely try to force an entrance within minutes of the premises being vacated, preferring to wait until the occupant is well clear. Even if they did they would almost certainly trip one of the other internal sensors. So the best course is to set the delays to give ample time to exit and enter, and control units having short maximum exit and entrance times are best avoided.

Another exit arrangement is the *exit set*. With this, when a sensor is activated on the exit circuit by the occupier leaving the premises after switching on, the whole system is thereby set. It is not primed until then, no matter how long the system has been switched on. So any delay on route poses no problem. Re-entry is by the usual delay.

An alternative method of exit is to use the regular sensor loop for the exit door but fit a key switch that bridges the door sensor in the door. When the door is shut and locked on leaving, the key switch is set to the open-circuit position, and when entering it is switched to the closed-circuit position.

This can work quite well with commercial installations but is less successful with domestic ones. The problem with these is that various members of the household enter the main exit door using their keys while the house is occupied and therefore while the alarm system is switched off. If they operate the key switch out of habit, it may be in the wrong position next time the system is switched on. Furthermore, it needs to be open-circuit when the system is switched on at night so that the door is protected, but closed-circuit when switching on during the day to permit exit from the premises. All this can generate much confusion and false alarms.

Business premises are normally unoccupied when locked up and so these problems do not arise. The bridging key switch is thus a viable alternative to the timed exit. The main risk is from forgetting to operate it, but as other keys have to be used at the same time to operate the locks this is unlikely. Apart from this, the snag is that of yet another key to use and carry around on one's keyring.

Another version of the key switch which overcomes most of the problems associated with it, is the lock-switch. This is a high-security mortice lock, that has a built-in switch that operates each time the lock is turned.

The lock is fitted in place of the existing one and so no extra key is required, nor can it be forgotten or left in the wrong position. It is probably the ideal solution to the exit and entry problem in certain situations. It could of course be fitted as an extra lock to the present one, and while this means an extra key, it also provides additional physical security.

The switch is usually of the single-pole changeover type which can be used to shunt the door sensor or to remotely set some types of alarm circuit. When it is used as a sensor shunt, the system must be switched on before leaving and off after entering, in the usual way.

With some alarm systems it is possible to switch the entire system on and off by means of the lock-switch on the exit door. A buzzer fitted near the door sounds continuously to warn of a protected door being left open, or other sensor being left in an actuated condition such as a chair standing on a pressure mat. There is thus no operation of the control panel at all.

This type of arrangement is ideal for premises such as church and public halls, for which there may be many parties having access with a key, but not all could be relied upon to operate an alarm system and carry out a proper exit routine.

Wiring to the lock switch is in this case vulnerable to tampering as the whole system could be disabled by it. A 24 hour anti-tamper loop run right to the lock is therefore essential, and the wiring should bridge the door-to-frame junction with a proper door loop consisting of multicore flexible wire in a sheath, terminated by enclosed junction blocks that have anti-tamper lids. Wiring from the lock to the loop should, if possible, be sunk in a groove in the door which is then covered with filler and painted over so that there is no observable trace. Otherwise it should be covered with metal channelling.

Lock-switch setting, though ideal for this situation, is not the answer in premises where different zones are required to be switched on independently due to staff working at different times in different sections of the building. It is only feasible when all zones are to be switched on at the same time, when the premises are locked up.

The bell circuit

The sounding device can be a bell or a siren, but usually it is a bell for reasons which we shall explore in a later chapter. It is obviously vulnerable being on the outside of the premises, and if silenced, the whole system is to no avail unless a communicator is incorporated in the system. The first essential then is to position the bell where it cannot be reached other than by a ladder. More than one bell on different sides of the building increases the security, as well as making more noise over a wider area. An internal bell is also highly desirable for the effect on any intruder who may somehow manage to breach the perimeter sensors and set off one of the inside ones.

Bells are commonly housed in steel boxes with anti-tamper switches under the lid connected to the 24 hour protection circuit. Any removal during the day will thereby set off the alarm. However, a possibility, which although unusual has happened, is for the box to be filled through the louvres with a foam sealant such as used by aluminium window installers, by bogus workmen or window cleaners. This takes but a few minutes from a pressure canister and effectively stifles the bell. Later they return for a much muted entry.

Anti-foam detectors are available which consist of an infra-red beam generator and receiver mounted inside the box. If the beam is interrupted by foam, the receiver triggers an alarm. However, with a rapidly filling box, the alarm is short-lived so protection is limited, and the system is little used.

A further disadvantage of the bell box is that the sound must escape through narrow louvres and is therefore reduced in volume. In particular the high frequency content on which the stridency depends is curtailed.

An alternative is to use a weatherproof underdome bell with a close-fitting dome, mounted without an enclosing box. Some of these are very difficult to silence mechanically, and they give a high sound output.

Wiring to the outside bell is likewise vulnerable and must be protected throughout its run. It should not be run along an outside wall, but come straight through a hole from the inside to the back of the bell. From there back to the control box it should be concealed and if possible buried in plaster, in tubing, and under floors. A 24 hour protection loop should be included in the wiring all the way to the bell, even if there is no box and anti-tamper switch.

If these conditions are met there is no way that the bell can be disabled without warning, if at all. However, it may not be possible for the wiring to be so protected over the whole of its run, or there may be some other factor that increases the vulnerability of the bell circuit. In such cases a self-activating bell or a self-activating module installed with a conventional bell will increase the security.

This device works with a rechargeable battery included in the bell box. It monitors the bell circuit and if the wire is cut it will sound the bell from the internal battery. Most self-activating circuits have a built-in charging facility, that keeps the battery fully charged, from a supply derived from the control box. If the alarm sounds without any damage to the wiring, power is drawn from the control unit in the usual way and not the battery. Thus the battery is conserved for emergencies.

3 Control equipment

The heart of any security system is the control box. Whatever facilities (or snags) the system possesses, they are determined by the control unit. Sensors and sounders are for the most part compatible with the majority of control boxes, so these can be ignored during the initial stage of planning a system. The first decision is what facilities are required, then a unit offering these can be sought.

In this chapter we will take a look at the facilities common to all units, and those which differ from one model to another. Some are intended only for domestic use, where the requirements are less stringent, mainly due to the reduced possibility of tampering, and the smaller area to be protected.

Cost and complexity

First, a word about cost. Control units have dropped sharply in cost due to the development of special semiconductor chips for security applications, and because of the increasing demand. Full-feature units now cost little more, or even less, than the basic models of a few years ago. If the job is carried out by an installation company, the majority of the cost will be for labour, as in addition to the actual work, it must be checked at all stages by a supervisor, and a comprehensive set of tests run at the completion, according to the National Supervisory Council for Intruder Alarms (NSCIA) code.

In relation to this cost, the difference between a complex control unit offering a large number of facilities, and a simple one with just the basics is negligible. In the choice of a unit then, cost can be discounted and the sole determining factor should be the facilities offered.

However, although cost-saving is not a factor, there is no point in having a box that bristles with features which you will never use. These added complications could reduce reliability, as the most reliable units are usually the simplest ones. Furthermore, a complex-looking control panel could create confusion among those having to operate it, especially in the case of non-technically minded staff. The chances of human error would thus be increased.

On the other hand, expansion or extension of the premises may require facilities not needed at present. So there should be careful consideration of all factors and possibilities before choosing a control unit.

Multi-zoning

The provision of a closed-circuit loop is common to all units as this is the principle mode of detection. Some are simple loops with separate anti-tamper circuits, while others have end-of-line resistors and use resistor-shunted contacts to eliminate the anti-tamper circuit.

There are differences in the number of individual zones that can be served. Unless zones are mentioned in the specification it can be assumed that the unit is a single-zone type and is intended for domestic use. Two and four zone boxes are common, while some are extendable by means of fitting extra printed circuit boards, to up to sixteen zones. This should be sufficient for all except the largest factory complex for which some models can be expanded up to 320 zones. However, several independent systems may be preferable.

Even in small business premises, it is usually advantageous to have two or more zones. As well as serving different areas, they can be used to take different types of sensor. Thus, movement detectors or vibration sensors can be connected to a different zone, than are the normal door and window switches. These can then be switched off in special circumstances such as stocktaking after hours, while leaving the perimeter sensors on. Also in the event of an alarm, actual or false, the precise circuit responsible can be quickly identified.

Zoned control units identify the zone which has been tripped, usually by means of an indicator lamp which remains latched on until reset. With some, if more than one zone has been tripped, an indication is given as to which was the first zone to be actuated. This may be done by the appropriate lamp flashing on and off, while those of other tripped zones remain permanently lit.

Annunciators

A useful device, which enables identification of tripped sensors with an existing single-zone unit, is the annunciator. In one model up to six sensors, or groups of sensors, can be connected with only four wires (six if an anti-tamper loop is included).

This trick is performed by a technical method known as multiplexing. Different value resistors supplied with the unit, are wired in series with each sensor, and when one is tripped the device recognizes that particular resistance value and a visible indication is given which latches on until reset. The effect is almost that of a multi-zone control unit, but some have no provision to actually control any of the sensors.

Normally-open-circuit loop

All units also accept normally-open-circuit sensors such as pressure mats. In the case of many domestic control units, the open-circuit loop is separate from the closed-circuit one. The practice with business systems is to combine them in the one circuit, a common method being to connect them between the loop and the anti-tamper circuit. With closed loops having an end-of-line resistor, the mats are shunted across the loop thereby shorting the resistor out when one is actuated.

Anti-tamper circuit

The anti-tamper loop active for 24 hour a day, is found on all business systems, except those having continually-monitored detection loops using an end-of-line resistor. Continuous loop monitoring or an anti-tamper circuit is necessary due to the high possibility of tampering during the time that the premises are open to casual callers. Some domestic systems have it but the low probability of daytime tampering renders it non-essential, unless there is a possibility of tampering by callers. Four wires must be run to all sensors when an anti-tamper loop is used, six wires for passive infra-red detectors. This is no problem as multi-cored cables are available with four, six, or more conductors.

Exit facilities

Exit circuits are provided in all systems but they differ in the way they operate. The various methods were discussed in the previous chapter. Most control units provide a closed delay loop that becomes active after the delay has expired. Provision is made to adjust the delay to suit the time required to exit. If an alternative method, such as a shunt switch on the exit sensor is preferred, the timed loop can be ignored.

Provision for the connection of a buzzer is commonly provided with most control units. This sounds during the exit and entry periods and with some models is also used for testing, to indicate a fault, or an anti-tamper circuit break. It can be mounted near to the unit and wired to it, but many units have a buzzer built-in. Alternatively, some have a built-in loudspeaker that generates one or more tones. Internal sounders save a little with the installation and are generally more convenient, but they have one snag. It is desirable to conceal the control unit, as much as possible, to save it from attack by intruders, as few boxes could withstand a hefty onslaught with a pickaxe or sledgehammer. But if intruders broke through the exit door and so started the entry timer, an internal buzzer could lead them right to the control box.

A buzzer mounted at some distance from the box, though audible at it and along the exit route, would confuse rather than aid intruders. Some units allow the internal buzzer to be disabled and an external one connected. Whether this is done depends on the level of security required. In most cases, the convenience of the internal one overrides the slight chance of it aiding an intruder, but for high-risk situations a separate buzzer is desirable.

Testing

All except the simplest domestic control units have provision for testing the detection circuits, but the extent of the tests vary. For the basic test, the system is activated, but with the internal buzzer or an indicator lamp switched in place of the alarm sounders. If there is a fault, that is a door left open, the loop broken or any other sensor is actuated, the fact is indicated without the alarm being sounded. In the case of a zoned system, an indicator shows which zone is at fault.

The system must then be switched off, the fault investigated and corrected, and then a further test made. If there is no fault some units have an ALL CLEAR indicator which shows that it is in order to switch on.

With some models, the main switch must be taken through a TEST position before it reaches ON. Thus it cannot be switched directly to ON without first checking for faults. However, the switch could be turned straight through the TEST position to ON and a fault indication ignored. If this is done with some units, the main sounders will not be activated, so preventing a false alarm, and the internal buzzer will continue to warn of a fault condition.

Other models have no separate test switch position. With these, the system automatically goes into a test mode when it is switched on, and any fault is shown by indicators or a buzzer, with the main sounders silenced. In some cases when the circuits are clear, all the fault indicators flash on to show that the circuits have been checked and then go out.

These tests merely check that no loop is open-circuit, they do not test individual sensors. Nor in most cases do they test the control unit itself; a 'no-fault' indication could be produced because the control circuits failed to detect one, and so would likewise fail to respond to an intruder activation. In the case of the door switches, most faults would show as an open-circuit loop. This is not the case with space protectors, which could fail without warning, or normally open-circuit detectors, such as pressure mats.

Although it is not feasible to do so at every switch-on, all sensors should be tested regularly, and this is accomplished by what is known as a 'walk test'. A provision is made in most control units to switch out the main sounders and substitute the buzzer, so that each sensor in question can be activated without raising an alarm. Each zone loop should also be checked by opening a door on that circuit. There is not usually any way of testing the 24 hour anti-tamper control circuitry other than by open-circuiting the loop to see if a fault indication appears. All this should be done during regular maintenance visits.

The bell is not usually tested, other than by an occasional sounding test. Some control panels have a provision for doing this. It is possible to silently test the bell by passing a small current through it and one panel which is no longer available did this at each switch-on. It appears that this facility is not offered with current control units, but some permit the sounder to be operated at low level.

The self-actuating circuitry in the bell box and the internal battery is not tested in any situation other than actuation by an open-circuit or shorted bell wiring. This is the weak point, and although required by the BS4737 standard, it seems preferable to take extra care in concealing and protecting the bell wiring over its whole length rather than using and relying on self-actuating circuitry.

System switching

The traditional method of switching the system is by means of a key-switch on the control box. Some manufacturers make 2000 different key profiles so the chances of the same one turning up in the same area are remote. Most key-switches have three positions although some have four. The positions usually are either OFF, TEST and ON; or ZONE 1, OFF, and ZONE 1 and 2. With the first one the switch must be taken through the test position to reach ON, while with the second, it is turned once either to the left or right to select the required coverage. With multizone units, separate switches which may be toggle or press buttons, select the zones to be activated. The main key-switch may then be inscribed PART ON, OFF, and FULL ON, so that a preselected group of zones may be conveniently switched in.

Keys are removable in any position, and most control box makers supply only two with no reordering of duplicates in order to maintain security. If more are required they can usually be cut locally from one of the originals, although this does compromise security.

Most people find keys a nuisance, especially when so many different ones have to be carried. With this in mind, many control boxes are operated by a key-pad similar to the buttons on a calculator. There are now about three key-pad models for every two key-switched ones. A three- or four-digit code is entered to set and switch the system off. A code may be already built-in to the unit, but in most cases it can be changed by the user to one of his own choice. Some models have both a key-pad and a key-switch, so that both or either can be used if required. A few use six-digit codes for extra security.

A three-digit code can offer 1000 possible numbers including 000, whereas a four-digit one offers 10,000. The latter is thus ten times more secure as it is ten times less likely for an intruder to hit on the correct code by chance. With some panels a digit cannot be used more than once in a particular code, so the possible numbers are reduced to 5040 for four-digit and 720 for three-digit codes.

It may be possible for an intruder to gain entry without setting off the alarm and then try to switch it off by experimenting with different codes on the key-pad. However, if the wrong code is entered with some panels the entrance timer starts, if it is not already running. This gives sufficient time before the alarm sounds, for the user to enter the right one if he mis-keyed the first time. Yet it would be insufficient for an intruder to try a long list of possible numbers.

However, he may have time to try a few possibles that he may deduce are likely combinations. So in choosing a code, care must be taken to avoid the obvious. Many users choose birthdates or anniversary dates, but these could be easily discovered by someone planning a burglary. Consecutive numbers forwards or backwards, should also be avoided although they may be easy to remember, as any such sequence could be covered by keying 0–9 and back again. A row of the same digit, especially 999, is another one to steer clear of.

There is no harm in choosing a number that is easy for *you* to remember because it has some other significance, but make sure it is one that no one else knows or could guess.

With some models the system goes into the alarm state if more than a certain number (thirteen in one model) of digits are keyed. This permits a second or third try after a mis-keyed entry of a four-digit code, but does not allow many random tries. The intruder would thus have to get it right by the third try, a very unlikely chance.

Some of the more sophisticated key-pad units respond to more than one code. One could be a cleaner's code that will set the system but not switch it off. Another, an engineer's code that enables routine testing to be carried out without divulging the main code. Thus, if any of these codes should fall into the wrong hands, security is not compromised.

A few models have a number of entry codes, each of which can be assigned to different employees. Should one leave the company, his code can then be deleted or changed without everyone else's being affected. With these a master code, when entered, allows any one of the others to be altered, so enabling a change to be made without delay. Another variation is to allow the system to be set by entering only the first two digits of the code. This can be more convenient when leaving the premises, and permits unauthorized persons to set the system without knowing the full code.

Some panels have a *duress* code, in which the last digit is different from the normal entry code. If entered under duress, it initiates a personal attack alarm. This may sound the full alarm, or it may signal to a distant point so giving the impression that the alarm system has been switched off. The hostage thereby avoids retaliatory violence, while alerting authority of the situation.

Though offering such facilities, many key-pads do have one disadvantage. These are the membrane type, that is, those that do not have actual press keys, but a touch-sensitive membrane on which the numbers are inscribed. They should be operated with the front of the finger, but as they are mounted vertically, most users poke them with their finger ends. They thus become indented by finger nails, especially from female users whose nails are usually longer.

This indentation can damage the membrane, and in addition betray the numbers of the code, although not their sequence. However, there are but forty possible combinations of four digits and only twenty-four if the same digits cannot be repeated, so the sequence would not take long to discover if the panel has no excess number limit. Another way of finding the four digits of the code is to dust the keys with chalk powder, whereupon it will stick to those having a greasy deposit from frequent touching. These ploys would be of little help though where several different codes are employed, as then most numbers would show signs of use.

A number of control boxes permit the connection of remote control consoles. These are small units that can be stationed at key positions within the protected area and allow all the functions of the master controls to be carried out. With one model up to four remote units can be installed.

Indicators

Some control units look formidable, with a row of lights to indicate various things, along with the control switches, a key-pad and sometimes a read-out screen. They are not quite so bad as they look. First there is the ubiquitous power lamp, which is on all the time that the power is connected. Apart from that most of the others are associated with one of the circuits: anti-tamper, exit, personal attack and one each for the zones. These come on when there is a fault in that particular circuit at switch-on or test.

In the event of an alarm, the light shows in which of the circuits the intrusion originated. In some cases they all come on momentarily at switch-on, to indicate that all are functioning and then go off. Some units have a 'clear' light which comes on when testing to show that all is well.

Some microprocessor controlled panels have a read-out screen which gives the status of the various detection circuits, prompts to take certain actions, and a record of the last alarm events with day and time. Some record up to 500 alarms, but if there were ever that number it would be better to move!

Output facilities

These vary between models. Most control units supply a maximum of 1 A to the sounder circuit, but some provide only half an amp (500 mA). It is recommended that normally the load be no more than 80 per cent of the maximum rating. As bells can take up to 300 mA and sirens from 0.5 A to 3 A, it is important to ensure that the control unit will supply sufficient current especially where two or more sounders are to be used. Also, it is necessary to choose sounders that draw less total current than that supplied by the control box.

All modern control units now have output timers that cut off the bell after a pre-selected period. The recommended time is 20 minutes. This has been introduced in response to widespread complaints by neighbours about bells sounding for long periods, before the owners have arrived to silence them. This is particularly annoying when triggered by a false alarm as most of them are. Other internal sounders that cause less annoyance may continue until switched off, as well as the external flashing light.

Many control units have a further facility, in that they contain, or will trigger, a communications device that dials 999 and give a recorded message, or one that has a direct line to the communication centre of a security firm or British Telecom.

In such cases, some control units enable the sounders on site to be delayed, to allow the police to catch the intruder. The wisdom of this is rather dubious though, as much damage could be done and the culprits depart with items of value, before the police arrive. Really, it is best to keep them out if at all possible, and the most practical way of doing that is to startle and scare them off.

Power supplies

The power for running the system is taken from the mains supply, with back-up batteries which switch in automatically if the supply fails. The control box must be permanently wired to a suitable mains connection box. It should never be supplied from a plug and socket as the socket could be switched off or the plug be pulled out. This could be done innocently, by someone such as a cleaner seeking a supply for the cleaning equipment, just as it could deliberately by an intruder.

The power supply is probably the weakest point in the whole system. A breakdown in the public supply, industrial action of the type seen in the 1970s, a circuit-breaker being tripped by a fault on the same house circuit or even deliberate sabotage of the supply, leaves the system at the mercy of the standby batteries. These have a limited capacity in the order of hours rather than days. The BS 4737 stipulates a minimum capacity of 8 hours which is woefully inadequate and is insufficient protection for even a weekend quite apart from longer holiday breaks when many burglars are most active. Most standby batteries exceed this capacity, but the time is still limited, being 2½ days on average for a typical control panel and rechargeable battery. Even this is reduced if passive infra-red detectors (PIRs) or other space protectors are used.

Some domestic systems use dry batteries that last a year or more, instead of the mains, and it may be wondered why industrial control gear cannot be made to do the same. The reason, is the extra current drain required by the various control circuits. The anti-tamper loop takes current as well as the detection loop, and furthermore is active for 24 hours. Zoning, which is not usually found with domestic systems, multiplies the loop current and requires current consuming indicators, while space protectors, the principal current consumers, are often used with commercial systems.

Dry batteries would have a very limited life supplying all these facilities and so are not practical for most business systems.

The length of time that a system will continue to function after the mains supply has failed, will thus depend on the control unit, the number and type of sensors, and the rated capacity of the battery. Strangely, it is a factor that is rarely mentioned in system specifications. It is assumed, it seems, that supply interruptions will be rare and short-lasting – an assumption that could prove dangerous!

Reset

Most units either have a reset button, to clear the alarm circuits after being triggered, or they reset automatically on switching off. In the case of those connected to a 999 dialler or other outside communicator, it is the normal practice for the user to be unable to reset the system which can only be done by an engineer.

The reason is to prevent false alarms. If the original alarm was false, due to some malfunction, merely resetting and switching on again without finding the trouble will only produce another false alarm and a cool reception from the local constabulary. When an engineer is called out after a false alarm, he makes a thorough check to find and clear the cause before resetting the system. The chance of a further false alarm is thereby greatly reduced.

When the bell timer has expired after an alarm, some systems reset but inhibit those detection circuits that are still in an alarm state; this stops them re-sounding the alarm, yet the rest of the premises continues protected. The facility is optional in many models and can be suppressed if desired.

There are now a large number of manufacturers each offering a range of units in a competitive market. This breeds innovative ideas and new or different solutions to old problems; it can also breed gimmicks. So it may be expected that units will appear offering something different, and this is good for progress. However, be wary of rushing into something new just because it is new, or that seems, or is claimed, to be better than anything ever produced before. Maybe it is, but also there may be serious snags that will only be discovered later and in certain situations. Many a manufacturer has launched a seemingly promising product only to find it a disaster and themselves bankrupt. Customers who bought it, were left high and dry with no spares, maintenance, or advice. Where security is involved it pays to be cautious.

4 Sensors

Sensors are a vital part of any alarm system. These are the guardians that respond to any disturbance caused by an intruder and trigger the main alarm circuit. Having done this, they play no further part in the operation, as the latching circuit in the control unit takes over to keep the alarm sounding. Hence any attack on the sensor, or its wiring, after the alarm has sounded is of no avail in silencing the system.

The sensors must be perfectly reliable and operate every time they are actuated, but they must also operate *only* when triggered by an intrusion and not generate false alarms as a result of wind, traffic, vibrations and other causes.

Switch contacts

Most sensors consist of a switch; either the contacts are *normally-open* (NO) and are closed when they are actuated, or they are *normally-closed* (NC) and are opened when operated. Some sensors have two pairs of contacts, one pair opens while the other closes, others have three contacts, one of which is common that switches between the other two. These are called *changeover* (CO) or *single-pole double-throw* (SPDT) switches.

Normally-closed contacts are used in continuous loop arrangements, where a current constantly circulates when the system is armed. Actuation opens the contacts and stops the current, which then triggers the alarm. This is the main method of detection, so chosen because cutting or breaking the wires also stops the current and sounds the alarm. There is thus a built-in security safeguard. This obviously cannot give warning of interference when the system is switched off during working hours, yet the possibility of tampering is greater then by casual visitors.

A second loop is therefore run to all sensors in commercial systems which carries current 24 hours a day, even when the system is not armed. This loop is not switched in any way by the sensors, it merely runs to terminals on them which serve as connectors. Any tampering with the wiring thereby gives an alarm indication at any time it occurs. Most sensors thus have at least four terminals, two to the switch, and two internally connected together for the anti-tamper loop.

Normally-open sensors trip the alarm when they are closed. These are mostly pressure mats which are difficult to make in a normally-closed mode. However, an anti-tamper loop is also run to these in commercial systems, so that they are protected against daytime interference.

The changeover type of sensor offers even higher security against tampering by using both normally-open and normally-closed operation for the same sensor. Actuation of either mode triggers the alarm as well as tampering with the anti-tamper wiring. These are not usually considered necessary, except for premises where very high security is required due to the possible attentions of determined and knowledgeable professional intruders.

Microswitches

The simplest type of switch is the microswitch (see Figure 7). As its name implies, it is a small switch which can easily be fitted in door or window frames, shop display cases, safes, cupboards, cabinets, drawers and under desk lids. It is commonly used in bell boxes and control units so that it is actuated if the cover is removed and breaks the anti-tamper loop.

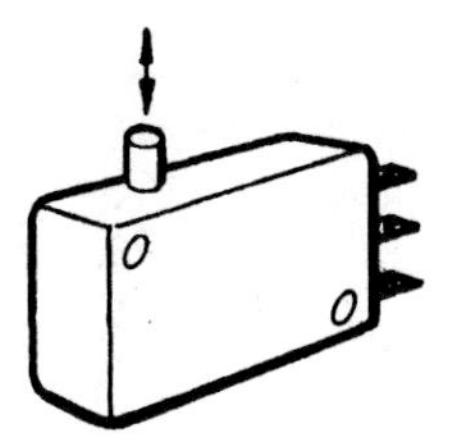

Figure 7 Basic microswitch. Arrows show direction of operation.

Unlike ordinary switches that operate by the up-and-down movement of a lever, the basic microswitch is actuated by a plunger which is sprung, so that it is depressed to operate and returns to its rest position when the pressure is removed. The amount of travel required by the plunger and pressure needed varies considerably from one switch to another, and selection is determined by the application.

Some makers give full mechanical specifications as to the various parameters of the plunger travel (Figure 8) and these are necessary for the selection of a switch for a particular purpose, especially when the movement involved is small. From the start position, the amount of plunger travel before the switch contacts operate is known as the *pre-travel*. Movement from this point onward is termed *over-travel*, and while a certain amount is

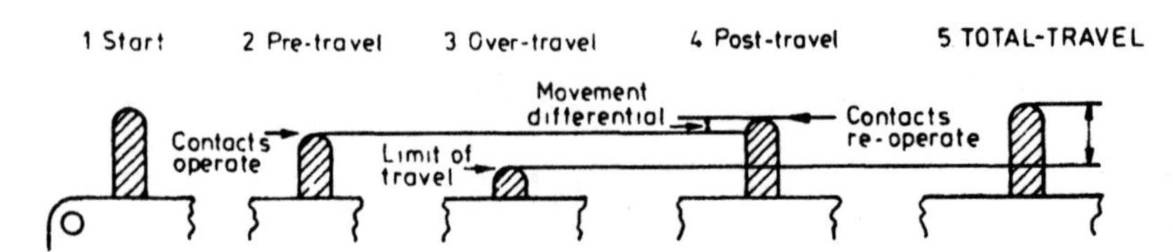

Figure 8 Microswitch operation chart showing operating points.

necessary to ensure that the contacts have in fact operated, it should not exceed the stipulated *limit-of-travel.*

On the release of the plunger, the travel from the limit to the point where the contacts are re-operated, is designated the *post-travel,* and it may be noted that the re-operation point is not exactly the same as the initial operating point, it being nearer to the rest position. The difference between the two is known as the *movement-differential.* The distance between the at-rest and limit-of-travel positions is logically described as the *total-travel.*

The forward stroke is of little interest for alarm applications as the sensor is normally held in the depressed position, but it is the return stroke which requires consideration. Post-travel must not be too small otherwise vibration and other disturbances may actuate it. Even a door with a loose catch may initiate a false alarm as it may not always be shut closely.

The operating movement of some microswitches is in the order of 0.3 mm which is obviously far too critical for normal door or window operation. These types are designed for much more sensitive applications. Long travel can be selected for doors or desk lids as these need to be opened wide to obtain access. This will reduce the possibility of false alarms. Shorter travel can be selected for more precise and critical operation such as safe doors.

An external attachment often supplements the basic microswitch to alter the mode of operation (Figure 9). There are three main types, the leaf or lever attachment, the wire actuator and the roller attachment.

The leaf or lever attachment is hinged at one end and passes over the plunger, so that movement at the free end actuates it. The effect is to amplify

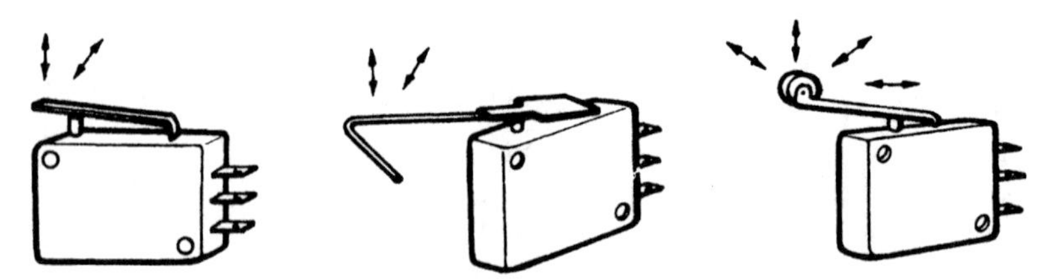

Figure 9 Microswitch attachments. Arrows show operating directions. (left) Leaf or lever actuator, (centre) wire actuator, (right) roller actuator.

the movement needed to operate the switch, and the longer the lever the greater the movement required. Various lengths from ¾ inch (18 mm) to 3 inches (76 mm) can be obtained.

Operation of the basic microswitch must only be by vertical pressure on the plunger, any movement with a sideways or wiping action could deform, jam or break it. Such movement is permissible with a lever, but it must be in one direction only, and that is away from the hinge for the downward stroke. Some makers have models with adjustable hinge positions so that leverage and travel can be varied accordingly.

The second attachment is also a lever, but instead of a flat blade it consists of a stiff wire with a right-angled bend at the free end. The operation is the same as for the bladed lever but the wire serves more as a feeler. A thin twine trip cord could be attached to the wire, or it can be used wherever a sensitive touch is required.

Neither of these can be used where a sliding contact is needed. For this, the roller type of attachment is necessary. Rollers are usually mounted at the end of levers, but they are also available fitted into the actual plunger. They have a particular application for sliding doors and drawers. The advantage is that they can be fitted an inch or so from the end of the frame so that they will not be actuated until the door or drawer end reaches that point.

If, then, the door is not fully slid home as often happens with sliding doors, it will not trigger a false alarm. Yet, if opened enough to gain access, the switch will be actuated. So both high security and immunity from false alarms can thus be achieved. It should be noted, that while for many applications, the magnetic switch to be described later is preferable to the microswitch, sliding surfaces are best served by roller microswitches.

Most microswitches have a pressure rating ranging from between 7 g and 454 g. Two figures are given, an operating pressure and a maintaining pressure. For most applications the pressure is not too critical, but for some delicate operations, the required pressure would need to be low. If too great it could actually hold off the actuating surface and prevent the switch operating.

Various fixing arrangements are possible as there are different fixing hole placements. Some have holes passing through the sides for sideways fitting, other have holes through the top plate for flush mounting, while others have a screwed bush through which the plunger passes for single-hole panel mounting.

In spite of their size, microswitches have high current ratings, but this is usually of little importance in latching alarm circuits as these pass very small currents. There may be some applications though, where the switch is required to operate an alarm bell or other load device directly. Magnetic switches cannot be used because of their low current ratings so the microswitch is the solution. Most are rated at 5 A which is more than adequate for most purposes, but some go up to 20 A.

Operating life is normally high and is usually specified by the maker. The lowest is some 100,000 operations, but typically it extends up to 10 million. These sorts of figures are of interest to designers of industrial counting or sensing equipment, but they would never be approached with normal alarm applications and so they can be ignored.

The majority of microswitches are made with single-pole double-throw (SPDT) contacts which thereby enables them to be connected in either the normally-open or normally-closed mode. They can also be incorporated in a security door lock to remote-switch an alarm system as this often requires SPDT contacts. An alternative that is sometimes used, is to house the microswitch in the door-frame cavity so that it is actuated by the lock bolt. This avoids the need to run wiring to the door itself which requires a special security loop connection across its hinge side. The disadvantage is that the microswitch needs careful adjustment so that it is fully actuated by the bolt, but is not driven beyond the travel limit.

Magnetic switches

A disadvantage of microswitches is that they are vulnerable to physical damage. The plunger or lever attachment can suffer deformation and jamming or even breakage by physical impact. Inaccurate positioning can cause the plunger to be driven beyond its limit-of-travel and so result in damage. If exposed in a door-frame or a similar position, it could be tampered with by sticking the plunger down with chewing gum or super glue. This could not be revealed by a normal test as the switch contacts are in their normal position. The damage would only come to light at the next walk test, which may be too late. All these are possibilities, unless the switch is well protected by being mounted invisibly or in a non-accessible position, such as inside a desk, cabinet or lock.

The problem arises because mechanical movement is required to actuate the switch and the actuating portion is open and exposed. Ideally, a sensor should have no external moving parts and should be completely protected by its surroundings. These characteristics are closely achieved in the magnetic reed switch. It consists of a pair of leaf contacts completely sealed inside a glass tube. Each contact is supported at opposite ends of the tube, and they overlap at the centre where the contact is made (Figure 10). When the device is located within an external magnetic field the contacts become temporarily magnetized and attracted to each other, and so they close. When the field is removed, they are demagnetized and spring apart.

The tube is encapsulated inside a plastic case which is mounted in a door or window frame. A matching case containing a bar magnet is fitted to the door or window so that when it is closed the magnet lies adjacent to the switch. The contacts are thus magnetized and close, so they can be

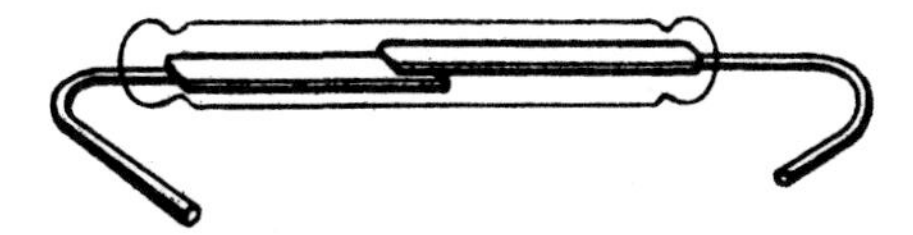

Figure 10 Reed magnetic switch. Reed contacts are sealed in a glass tube and are actuated by an external magnetic field.

connected in series with a closed loop. When the door or window opens the magnet moves away and the contacts open.

We should here clear up what may seem a confusing description, the device is described as *normally-closed,* because when it is installed and the system is on guard with the door shut, the contacts are closed. This of course is because they lie in the magnetic field of the mating piece. However, when the switch is free, before it is installed, the contacts are *normally-open,* because they are not influenced by any magnetic field. Care must be taken over this when installing, because the switch contacts will show open-circuit if checked with a continuity tester, whereas the linked anti-tamper contacts show closed. It is easy to be confused by this and connect up the wrong pair.

The magnetic reed switch offers numerous advantages over the microswitch for protecting doors and windows. Having no external moving parts it is less vulnerable to damage, and is more difficult to defeat. It does not rely on actual contact with the door or window, and so is not affected by vibration or minor irregularities such as wood swelling or contracting. There are no restrictions as to the angle of travel of the magnet either toward or away from the switch. The only possible disadvantage is that two units must be installed instead of one.

Positioning and distance between the units is not as critical as with the microswitch and its actuating surface. In fact there can be quite a gap between the door and its frame without causing problems, a distinct advantage with many properties.

The distance at which the magnet will influence the switch, varies according to the type of switch and strength of the magnet. Two distances are quoted by the makers: *operating-distance* and *release-distance.* The former is the distance at which the approaching magnet will cause the contacts to close, and the latter the distance at which the receding magnet will allow them to open. The release-distance is roughly one-and-a-half times the operating-distance, but in some cases can be twice as far. For door and window sensors it is the release-distance which is the significant parameter.

Typical release-distance can be from ⅜ inch (10 mm) to 1½ inches (35 mm). These represent the distances that the door or window will open before the alarm is triggered, assuming that the units were in close contact with each other to start with. If there was a gap, which is virtually certain, the gap would have to be subtracted from the release-distance to find the opening distance at which the alarm would sound. The gap must always be less than the operating-distance, otherwise the contacts would never close.

Distances quoted by the makers are subject to a certain tolerance between individual units and are measured with a new magnet. As magnetism is lost with age, distances will be reduced as time passes. To ensure reliable operation at all times and with all units of a particular type, minimum distances should be reduced by 25 per cent.

Switch contacts are of precious metal and being sealed in are unaffected by atmospheric moisture or pollution. They have therefore an extremely long life, over 100 million operations, which is ten times that of the average microswitch. So if a door was opened and shut once every 10 seconds during an 8 hour working day, 7 days a week, the contacts would last for over 95 years.

Although such a life expectancy is never likely to be reached, it gives some idea of the inherent reliability of the magnetic reed switch, and another reason why it is so eminently suitable as an alarm system sensor.

As with the microswitch, some reed switches are available with SPDT changeover contacts so they can be used in the normally-open as well or instead of the normally-closed mode. These are not often used except for systems requiring the highest possible security.

The current rating is limited and the magnetic switch should not be used to directly drive a bell or other signalling device. If such a circuit is required a microswitch should be employed.

Encapsulation

There are several different types of encapsulation to suit various mounting requirements. One is the circular flush-fitting type, which is inserted into a hole drilled in the door frame, a flat disc forming a head which lies flush with the surface. The magnet is fitted into a matching unit that is similarly sunk into the edge of the door (Figure 11).

The result is inconspicuous and the only woodwork needed is the drilling of the holes to accommodate the units with another smaller hole meeting the bottom of the switch hole to take the wiring. As the units are about 1½ inches (35 mm) long, there must be at least this depth of wood to receive them. This

Figure 11 Circular magnetic switch and matching magnet.

Figure 12 Flush-mounted shallow rectangular magnetic switch and matching magnet.

may not be the case with shallow frames or for windows or glazed doors. Another factor, is that as the mating surface area is small, the switch and magnet must be accurately aligned with very little tolerance for misalignment.

Another type for flush fitting is the shallow oblong variety (Figure 12). As these are only a few millimetres deep they can be installed where the wood is not very thick such as in window frames. A cavity must be chiselled out to accommodate them and they are supported by screws through the overlapping faceplate. The faceplate is normally proud of the wood surface, but if a really unobtrusive job is desired this too can be recessed into the woodwork. If puttied in and painted over, it can be indistinguishable. As the mating area is quite large, accuracy in lining up the switch and the magnet is not so important.

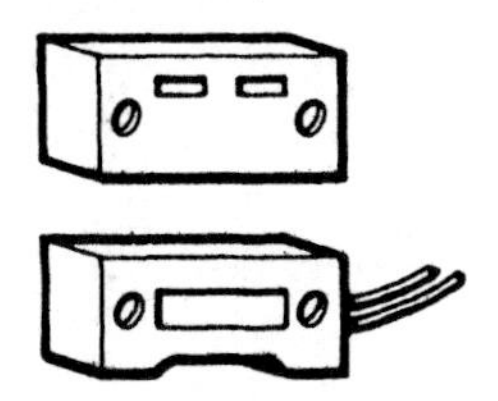

Figure 13 Surface mounted magnetic switch and magnet.

There is also an oblong surface-fitting type as shown in Figure 13. These are mounted on the surface of the frame, on the opening side, and the magnet on the door surface, so that they meet edge-to-edge. Being visible and accessible these are much less secure than the flush-mounting varieties. They are sometimes supplied with DIY home kits to make fitting easier for the amateur, by saving the drilling involved with the flush mounting types. As they are mounted on the inward side of the door which opens into the protected area, they cannot be tampered with from the outside, so the only possible interference would be from the inside, an unlikely event in a domestic system.

With business premises no such assumption can be made and the vulnerability to tampering of these devices render them quite unsuitable. Where there are metal windows, there may be no alternative, as it is virtually impossible to fit a flush-mounting type with these, unless as with some double-glazed aluminium windows, there is sufficient room in the hollow section from which they are constructed to accommodate a magnet and a switch in the frame.

If a surface type cannot be avoided, it should be painted over to look as inconspicuous as possible, and backed up with pressure mats or space protection devices inside.

Although unusual, it is not impossible to defeat a reed switch by using an external magnet. First of all the intruder needs to know its position in the door frame. This can easily be discovered by running a compass around the frame. The magnetic field from the associated magnet will turn the compass needle and give an accurate indication of the switch location.

Now, all he needs to do, is to use a strong external magnet to immerse the reed switch in its magnetic field. This can be fixed temporarily to the door frame in some way, and then the door can be forced open. The magnet thus holds the reed switch closed and no alarm is triggered.

This method is unlikely to be used by the average intruder, because it requires knowledge and some skill instead of brawn, but it could well be employed by the professional thief. If the premises contains goods likely to be attractive to the professional, this possibility cannot be discounted.

One method of increasing security against this trick, is to fit two switches to each perimeter door frame, one high and the other very low. On discovering one, possibly the high one, the intruder could likely assume it was the only one and not search further, especially right down to floor level. He would thus be caught by the one he missed. If he did find the two, the chances are he would only have one magnet and so would be thwarted anyway.

Another way to reduce the possibility of defeating the magnetic switch by use of an external magnet, is to use a high security type which contains two reed contacts that need energizing by both poles of a magnet (Figure 14). The mating magnet is cylindrical but is bifurcated by a central slot with opposite poles being formed in each section. These sections must be lined up to the two sets of reeds in the switch when installing in order to actuate them. The use of an external magnet or magnetized strip will not operate the switch and so it cannot be neutralized by this means. This type of sensor is recommended for installations where the highest security is required.

It should be noted that with all magnetic switches the close proximity of ferrous metal, such as metal window frames, may affect the operation, particularly the operating distances.

Figure 14 High-security double-reed switch. Can only be actuated by both poles of a magnet, hence the bifurcated magnet shape.

Connections are of two types, those having terminals and those with lead-out wires. The terminal type are the easiest to fit and are quite satisfactory, providing the terminals are well tightened. The lead-out wires of the other type must be soldered to the circuit wiring and although less convenient to install they ensure a permanent joint that will not work loose and cause baffling intermittent false alarms. Some are obtainable with 3 ft (1 m) of lead-out wires and are designed for aluminium or uPVC windows, being housed in either aluminium or white plastic casings. The long wires enable a joint to be made unobtrusively away from the window frame.

Some applications need a particularly robust unit that will stand up to adverse conditions. Roller shutters, up-and-over garage doors and bank shutters call for specially strong encapsulation. Switches are available that are sealed in aluminium cases that are flat on one side, but pod-shaped on the other (Figure 15). The depth is a mere ½ inch (13 mm), and the unit is designed to be screwed to the floor, using plugs if the floor is concrete.

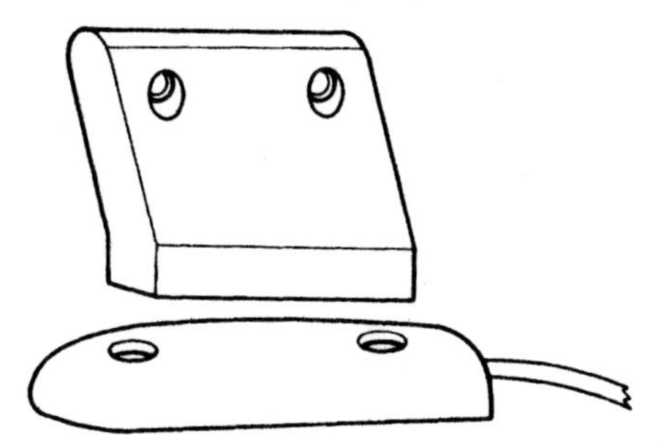

Figure 15 Garage door switch with magnetic actuator.

The switch can be driven over by a car, without damage, and there is one model which it is claimed will not suffer if driven over by a fork-lift truck! The magnet is housed in an aluminium casing that is fixed to the door, and is angled inside the case to minimize the effect of a steel door.

Diode-sensor muting

A novel method of protecting the loop circuit as an alternative to the separate 24 hour protection loop, is diode muting. A diode is a device which allows a current to pass in one direction but not the other, and each sensor has a diode connected across it (Figure 16).

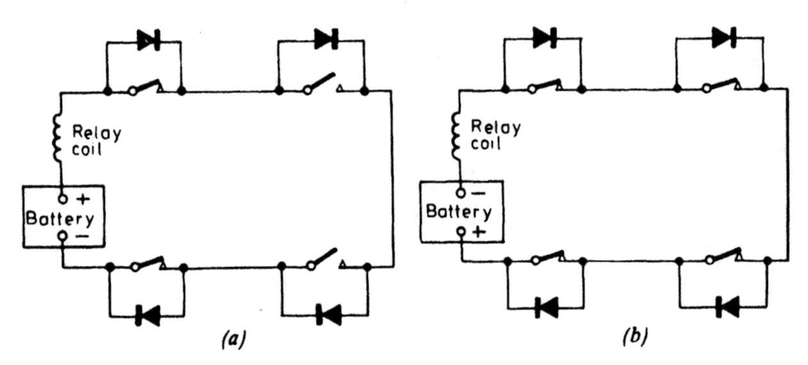

Figure 16 (a) Diode muting of loop sensors for day-time monitoring. Polarity of battery is such that diodes conduct and so the loop is continuous whether sensor contacts are open or closed. Cutting the wiring breaks the loop and sounds the day alarm immediately. (b) Battery polarity is reversed for night alarm service. Diodes no longer conduct and loop functions normally.

For daytime monitoring of the loop, the current supplied by the control box is reversed in polarity. Current flows through the diodes which effectively short-circuit each sensor. Thus, a loop current is maintained even though protected doors are being opened and closed. Only if the wiring is damaged will the current cease and an alarm state be triggered in the control box.

When the system is switched on, the loop current polarity reverts to normal. Now, the diodes do not conduct and have no effect, so any actuation of a sensor breaks the loop continuity and sounds the alarm. A feature of the system is that one or more unbridged sensors can be connected into the loop. These will then give an alarm indication when operated day or night and could be used to protect especially sensitive areas. Legitimate callers would have to operate a bypass switch with a pass-key to gain admittance.

As the loop could be defeated by short-circuiting it, as in fact any loop could, provision can be made to fit a terminating resistor at the furthest point (Figure 17). Then, the control circuit senses the value of the resistor and triggers an alarm if the circuit resistance falls below it (as a result of a short-circuit) or rises above it (due to an open-circuit).

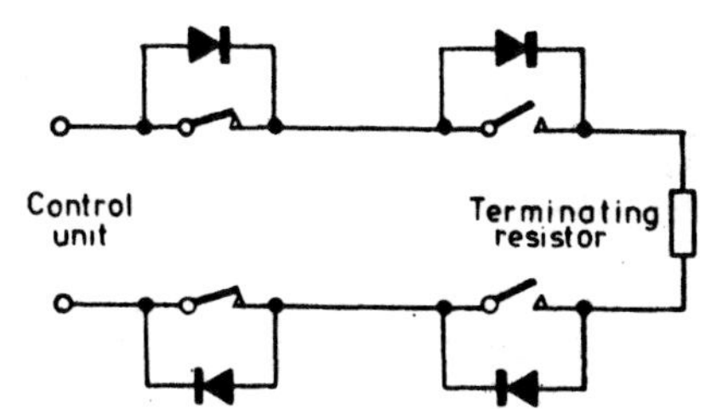

Figure 17 Diode-muted loop using terminating resistor. This foils attempts at bridging across loop as alarm is triggered for both open- and short-circuit conditions (Monive System).

There is a limit to the number of diode muted sensors that can be used on any one zone. This is because each diode drops a small voltage (0.6 V for silicon, 0.2 V for germanium diodes) when in the forward conducting state. Too many sensors could thus add up to a voltage greater than that applied to the loop so no current could flow in any direction when in the 24 hour monitor mode.

It may be wondered how the presence of the diodes affect the potential reliability. Diodes are generally reliable devices, but those not carrying a heavy current, such as power diodes tend to go open-circuit rather than short-circuit if they fail. So if a failure occurred the sensor would most likely become unmuted. This would not be a threat to security but merely trigger the tamper alarm when actuated during the day.

Another arrangement is to use a loop with a terminating resistor, but having normally-open sensors connected across it. Each one has a diode in series with it (Figure 18), but none of the diodes are in series with each other so there is no limit to the number that can be used. The diodes are connected to be non-conducting during the day and so they disable the associated sensor. When the system is switched on, the polarity is reversed, and the sensor conducts when actuated.

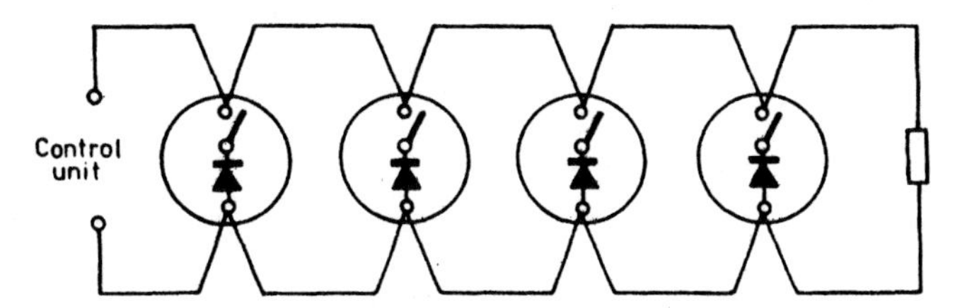

Figure 18 Resistor-terminated loop using parallel normally-open sensors. Muting diodes are in series with each contact but not in series with each other. This enables unlimited number of diode-muted sensors to be used in the loop (Monive System).

Normally-open and normally-closed sensors could be mixed in this loop providing the diodes were in parallel with the normally-closed sensors and in series with the normally-open sensors. The normally-open ones could be incapacitated by an open-circuit diode, and this would only be revealed by a walk-test. While diodes are generally reliable, this does reduce the level of security by a degree or so.

Resistor-bridged contacts

A similar system to diode muting uses a resistor across each pair of contacts in place of the diodes. An end-of-line resistor terminates the loop. When in the 'day' mode, a current circulates around the loop, monitoring that it is still intact. If contacts open when doors are opened, the resistor maintains the continuity. When switched to the night mode, all contacts are closed and the end-of-line resistor is the only resistance in the circuit. The control panel senses its value, so if a contact is opened and the resistance changes, an alarm is triggered (see Figure 6).

Contacts are available with a resistor built-in for use with this system, which can thus dispense with the separate anti-tamper loop and therefore requires only two wires.

Radio sensors

Another system, is the radio-controlled alarm system, consisting of sensors and a control unit. The sensors are conventional switches which can be surface mounted to doors and windows, but instead of being wired to the control unit, each sensor has a tiny radio transmitter powered by its own internal battery. When the sensor is operated, a radio signal is transmitted that is received by the control unit, which then activates and latches an alarm in the usual way. A feature of the system is that having no wiring, it is portable, and so can be quickly installed in temporary premises and just as quickly removed. The disadvantage is that there is no anti-tamper system and only limited monitoring is possible, due to the small internal battery capacity.

Transmitters are available that can be fitted to existing space protection devices, and receivers can be fitted to ordinary control panels. Thus a system can be part radio (termed free-wired), and part conventionally wired (called hard-wired). Sensors that pose difficult wiring problems, such as to outhouses or to odd remote windows or doors, could be of the radio type and so eliminate the wiring. Range is up to 200 ft (60 m), but this can be extended by an aerial.

Another possible use is in goods yards where valuable articles are being moved around. The trucks could be fitted with radio sensors, that could be inactivated only by a key. Any attempt at hi-jacking or theft when unattended, would thus be signalled to the security office.

A further useful feature of the radio system is the provision of radio panic buttons. These can be used wherever personnel are temporarily stationed or even when travelling within the premises. The range is the same as for the sensors and they can be carried inside a pocket or handbag. While a radio system would not replace a permanently wired installation, it could be very useful in the situations described.

Pressure mats

These are rectangular flexible mats that come in a variety of sizes but usually with an area of between 2–6 ft^2 (0.2–0.3 m^2). The usual method of construction is a sandwich of two sheets of metal foil separated by a layer of perforated foam plastic. When pressure is applied to the mat the foil sheets make contact through the foam perforations, and when it is released the foam separates them again. A plastic covering seals the mat against dirt or moisture. Care must be taken when installing not to puncture the cover, as although it may have no immediate effect, it could cause trouble later.

There are four lead-out wires which must be soldered to the circuit wiring. Two go to the foil sheets and are connected to the normally-open detection circuit, while the others are a straight-through link which connect to the anti-tamper circuit. To identify them, the switch wires are usually stripped for a few millimetres, while the anti-tamper wires are not.

Thickness is typically about ¼ inch (6 mm) and the mat should be installed under a carpet and underlay. A slight bulge may result, but this will soon bed down and become almost imperceptible after a while. In domestic systems, a pressure mat is often placed on a stair, but this may not be practicable with a business installation, because few stairways in business premises are carpeted.

Positioning is all important, as a mat placed where an intruder is unlikely to walk is of little value. It must be put in the most likely place he will step. This includes the area around valuable items, cash-tills, safes, chemist's drugs cabinets and so on. In domestic systems the positions to place them would be in front of video recorders, TV sets, hi-fi equipment, and bedroom dressing tables where jewellery may be expected.

The pressure mat can serve areas where perimeter protection is difficult or of a low order of security, such as where surface magnetic switches have to be used on metal windows. Really though, they should be regarded as a second line of defence. If intruders do manage an entrance without actuating a perimeter detector, they may trigger an internal sensor such as a pressure mat. But these should not be relied on, the principal object of the system is to protect the perimeter and keep intruders out.

If there is no carpeting at all as would be the case in warehouses and many offices and shops, pressure mats are unusable, but there are other devices known as space protectors that can be employed. In premises that are carpeted, even those that have space protectors, pressure mats should be used for inexpensive and basically foolproof back-up protection.

The amount of pressure needed to actuate a mat is rarely quoted by the makers. It is usually about 2.5–3.0 lb per in^2 (176–211 kg per cm^2). The average areas of a man's shoe sole and heel is about 25 in^2 (161 mm^2) which for a 10 stone weight gives a pressure of 5.6 lb per in^2 (395 kg per cm^2). A woman is on average lighter than a man, but the shoe area is much less, so the pressure is actually greater. A child's weight too is less, but so is its shoe area.

These figures assume that the foot is laid gently and evenly on the mat, but in walking this is not the case; the heel comes down first with some momentum, and so the force acting on the mat is much greater. There is little fear then, that it will not respond when trodden on. The weight of a cat is insufficient to actuate it unless it jumps down on to the mat, so the resident mouser is unlikely to be the cause of false alarms.

One thing that must be watched, though, is that objects such as stock, chairs, desks or other items are not inadvertently left standing on the mat. After a mat has been installed for some while the tendency is to forget it, and so there is a strong possibility of something being moved on to it and left there. The weight of the offending article may be insufficient to immediately create a contact, but it could settle after a few hours and trigger the alarm – in the middle of the night! This has been the cause of many a false alarm.

A drawback with pressure mats, is that being normally-open devices they cannot be part of a closed loop, and so cannot be tested along with the other sensors each time the system is switched on. They can only be checked by a walk-test, that is actually operating the device with the control unit in the armed or test mode.

Connection was once made to a separate circuit with its own wiring, but now the normally-open sensors are usually connected from the normally-closed loop to the anti-tamper loop, or across an end-of-line resistor when used.

Use only the best quality pressure mats. Cheap ones have inferior foam insulation between the foil sheets, which breaks up after a while, especially if located in a well-trodden position. If the carpet is thin or there is no underlay, the pressure mat will not last long. Even if well protected, be prepared to replace them from time to time.

Window strip

Various methods can be used for protecting window glass. One of the most common is strips of metal foil fixed across the inside surface of the glass. These are usually made of aluminium but lead is available for use in areas where high atmospheric pollution could corrode the aluminium. The strips are supplied in rolls of various widths, ⅛ in, ¼ in, and ⅜ in (3, 6 and 8 mm) being the most common. They can be either self-adhesive or non-adhesive, but the self-adhesive is the most convenient for normal applications.

The foil is run along a vulnerable area of glass and is connected into the closed loop circuit, the 24 hour panic button circuit, or an anti-tamper loop. If the glass is broken, the foil is severed and the loop open-circuited. For extra security a two-pole arrangement such as shown in Figure 19 should be used. The main run is connected to a 24 hour loop that is normally used for panic buttons, and the extra strip is taken to the anti-tamper loop. Any attempt to either cut the foil or bridge the contacts will produce an alarm condition, sounding an internal alarm when the system is switched off during business hours, or a full alarm when the system is switched on.

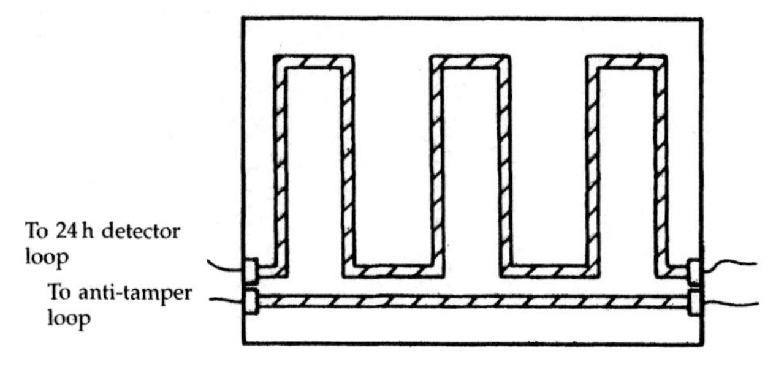

Figure 19 Window foil with connections to both 24 h detection loop and anti-tamper circuit.

Special adhesive terminal blocks are fixed to the glass at the side of the pane. These make contact with the foil strip and have terminal screws for connecting the wiring. A making-off strip is available for bridging two panes across a window frame.

Ideally, the strip should cover all parts of the glass to give maximum protection, but this may obscure the window display. An alternative, where lower security is acceptable, is to run two strips at roughly a quarter and three-quarter positions up the glass. This leaves the centre clear but gives adequate protection and can even be an embellishment. A further advantage is that the circuit wires can be connected at the same side of the window, the free ends being linked on the opposite side by a wire up the frame.

One feature of metal foil on glass is that it can be seen from the outside. An intending intruder is thus informed that the glass is protected by an alarm and so will be deterred from attempting an entry by breaking it. This is far preferable to triggering the alarm by some invisible detector after a breakage, as it saves the not inconsiderable cost and inconvenience of replacing the window. Deterrence is always the best form of defence.

There are obvious problems associated with fitting metal foil to windows that open. The connecting block would have to be fitted to the window on the hinge side, and flexible leads bridged over to another block on the frame to connect the circuit wiring. It is better then, to use a magnetic sensor for all opening windows.

While in most cases metal foil offers a good protection to fixed glass it is not unassailable. A determined intruder could use a glass cutter to cut a hole avoiding the metal strips. With a showcase containing small valuables, such as jewellery, the thief could then abstract the contents through the hole.

One remedy would be to run the foil strips closer together, but this would reduce visibility to an unacceptable degree. An alternative is to use wired glass as shown in Figure 20. This is a special glass consisting of two sheets sealed together with a series of silver wires across almost the complete width. The wire is very fine and is hardly visible, but being spaced about 2 in (50 mm) apart gives a high degree of protection. As with the foil, the wire is connected to the closed or 24 hour loop. This is obviously more expensive than the use of foil which can be fitted to existing glass, but it affords high security with good visibility where that is required.

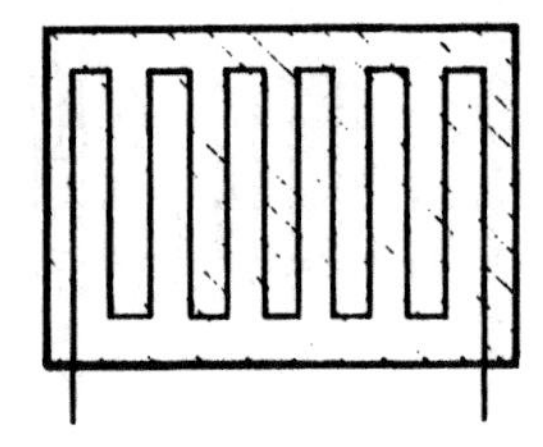

Figure 20 Wired glass consisting of two bonded sheets with silvered wire at 2 in (50 mm) spacing.

Ceilings

The same principle can be used to protect ceilings. These are often overlooked as a means of entry, but especially with single storey buildings, they can afford an easy way in for an intruder. Tiles can be quickly lifted off, or with flat roofs, the bitumous-felt covering can be soon penetrated. Next comes the heat insulation which is no problem to remove, and then the ceiling, which is easily broken through. The exit poses little difficulty as there is usually a door or window that can be opened from the inside. Even if this sets off the alarm, the thief is on his way out and is well away by the time the police are informed and arrive.

So, ceilings can and should be protected, when in a vulnerable situation. This can be done by running a fine wire to and fro across the ceiling then concealing and protecting it by papering or skimming with a ceiling finish. Even painting would be sufficient to retain such a wire in position and protect it.

Special hard drawn lacing wire is made for this purpose, or for similarly protecting door or partition panels. It is very brittle and snaps easily, thus giving excellent protection, but it needs care in handling and must be protected by some form of covering.

As an alternative, fine copper wire, varnish-coated to prevent corrosion, can be used, the type used for transformer windings being particularly suitable. Although less brittle than the lacing wire, it breaks easily enough when subject to stress if of sufficiently fine gauge. As copper has a degree of elasticity, it should not break if subject to slight movements due to structural settling. Providing the connections are sound there is little possibility of false alarms.

Acoustic detectors

Also called *sonic detectors*, these are not to be confused with the ultra-sonic devices described in the next chapter. The term embraces a range of sensors that operate mostly by the sound generated by breaking glass. Examples of the device are shown in Figure 21.

They consist of a microphone, amplifier and output relay. A special filter circuit is included which passes only those sound frequencies generated by breaking glass, and thus confers an immunity from false alarms caused by other sounds. One type, also termed piezo-electric because the microphone is of the crystal contact variety, is mounted on the glass by means of an

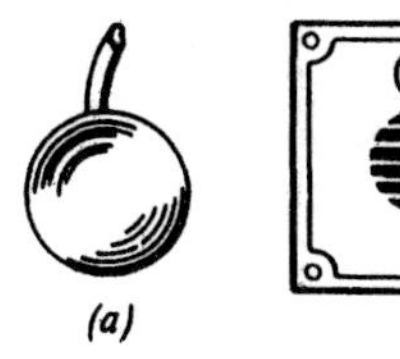

Figure 21 (a) Small acoustic detector for fixing to glass. Responds only to sounds of breaking glass. (b) Non-contact acoustic detector detects sounds of breaking glass over 15 ft (4.5 m).

adhesive. The complete unit can be very small, little larger than an inch in diameter. The filter eliminates low frequencies, passing those in the high 6000–8000 Hz range (piano top note is 4000 Hz).

Others are mounted on an adjacent wall or ceiling and have a normal air pressure microphone. Some of these have a more complex filter that responds first to certain low frequencies which are produced on first impact followed by the high frequencies generated as the glass shatters. They will not trigger unless both the required frequencies occur in the correct sequence, thereby giving an even greater immunity against false alarms. This type should not be used for wired, laminated or toughened glass however, as they do not produce the correct frequency pattern.

The glass-mounted type must be positioned at least 2 or 3 in from any window frame because the intensity of vibrations in glass reduces from maximum at the centre of the pane to zero at the frame. Both types protect an approximately 10 ft (3 m) radius of glass. Some have an indicator light which latches on to show if it has been activated, or which one if there are several. If glass has been broken this may seem to be superfluous, but it does help to identify the source of a false alarm if one should occur.

They offer an advantage over window foil for large areas where a lot of foil may be considered to be visually detracting, and especially for multipane windows where there may be practical problems in foil laying. The disadvantage is that the visual deterrent effect of foil is lost. For multipane windows the space sensor rather than the contact sensor should be used, as each pane would require its own contact detector.

A similar principle is used in wall sound detectors. With these, either structure or airborne sounds are picked up by the microphone and filtered so that only those produced by hammering, drilling and cutting activate the unit. Again it is mainly the high frequencies that are detected, so those produced by bumps, traffic noise and other normal happenings are ignored. A number of microphones can be connected to a single detector, so that a large area or several separate areas can be monitored. The detector thus processes the output from all the microphones and open-circuits the loop if any are actuated.

As all acoustic detectors contain an amplifier, they require a power supply, and this is provided by the 12 V auxiliary supply available in the control unit. Current consumption is 20–30 mA. An extra pair of wires is needed to convey this to the detector, so to include the anti-tamper loop and the detection loop, a six-wire cable is required.

Vibration detectors

These are comparatively simple devices consisting of a leaf spring suspended at its top and having a weight fixed to its free end. Also at the free end is a contact that mates with another which is fixed to the case (Figure 22). The pressure between the contacts that are normally closed, is adjusted by a set-screw which thereby sets the sensitivity of the device.

Any vibration or movement causes the contacts to part and so initiate an alarm. It can thus be used to protect windows as an alternative to the contact acoustic detector, or any wall or partition where forced intrusion is possible.

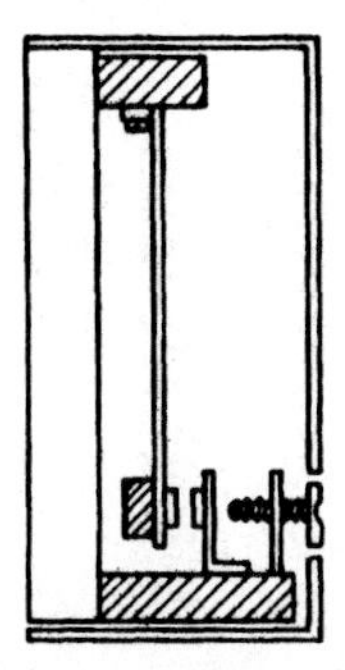

Figure 22 Basic principle of vibration detector. Pendulum contact in close proximity to fixed contact. The latter is moveable by screw to give adjustment of sensitivity.

The main disadvantage is the difficulty of setting the sensitivity. If set too high, it may be affected by random vibrations. Passing heavy lorries in particular, can make a large window vibrate strongly. There is no filter to eliminate inoffensive vibrations as there is with the acoustic detector. However, control boxes are designed to respond only to loop breaks longer than 0.2 seconds, so this helps to eliminate small vibrations which may just cause the contacts to part momentarily.

One advantage over the acoustic detector is that as plate glass is quite robust it may take several blows to break it. The acoustic detector is actuated only when the glass is broken, but the vibration sensor would almost certainly be triggered from the vibration of the first blow. With the alarm ringing after an unsuccessful first onslaught, the attacker would very likely be deterred from a further attempt, thus saving loss and damage.

Furthermore the vibration detector does not need a power supply, is cheaper, and requires less complex wiring. A large number could be used if required on walls, partitions or anywhere likely to be broken through. Its susceptibility to normal vibrations though, would preclude its use in busy areas where there is much activity, and especially on main roads.

Impact detectors

Impact detectors respond to vibrations or impact as do the vibration detectors but they are more sophisticated. A piezo-electric (crystal) rod is supported at one end and any vibration causes the free end to oscillate and thus generate a voltage in it. A built-in circuit analyses the nature of the signal thus produced and actuates a relay if it exceeds a pre-set level and duration.

The device is thus less prone to false alarms than the simple vibration detector, and as it does not depend on gravity acting upon a weight, it can be mounted in any plane. It requires a 12 V supply from the control box and is larger and more expensive than the vibration sensor. As each is self-contained with its own electronics, installing a number of them could be costly. It is therefore most suited for situations where only one or two are required. If more sensors are needed, it is more economical to use inertia detectors.

Inertia detectors

These sense disturbances and vibrations, but are particularly sensitive to low frequencies. Thus a gentle prising or levering of a structural member, or motion generated by climbing a perimeter fence which may be ignored by a vibration or impact sensor, will trigger an inertia detector. As its name implies, the mobile component has a high mass to give it inertia, so that when the environment is disturbed, the housing moves but the component does not. Contacts are thereby broken thus activating the alarm.

In one model, the component is a gold-plated ball seated on a pair of contacts so forming a normally-closed switch. Displacement of the ball from either contact breaks the circuit. A high security version, has in addition, a closely-spaced ring around the ball (Figure 23(a)). This serves as a normally-open switch, so that any disturbance causes the ball to make contact with the ring, as well as break the circuit across its supporting contacts.

The maximum current that can be switched by this device is low because the area of the ball actually in contact with its supports is minute. A large current passed through such a small area could cause pitting and even welding to occur. Specified current is 0.2 mA, with an applied voltage not exceeding 2 V. This is much lower than the 500 mA or so that the vibration sensor with its much larger contact area will permit. It is also lower than the current applied by many control units to the loop circuit. It thus needs its own control unit which will not exceed the maximum current, as well as process the result to distinguish between false alarms and intrusions.

With another model, a lever having a weight at its free end is supported by a gold-plated rod which rests on a contact piece. An angled extension of

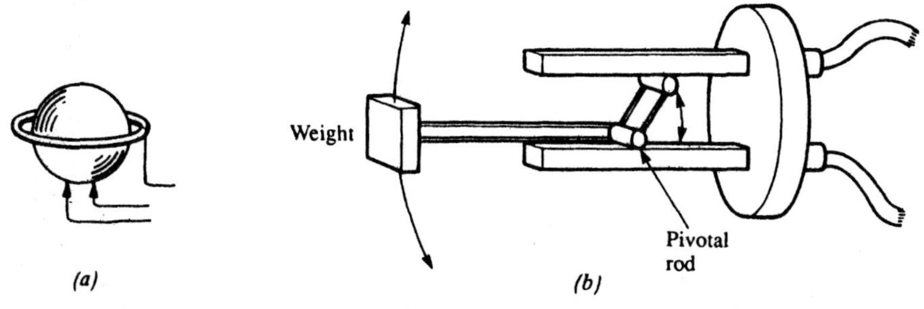

Figure 23 (a) Inertia detector. A ball rests on two contacts forming closed circuit. An extra ring contact surrounds the ball. It is used for detecting motion and vibration. (b) Inertia detector having a weight on the end of an arm which pivots on a lower contact when jolted, thereby breaking the contact against the upper contact.

the lever is terminated by another gold-plated rod that rests against the under surface of a second contact piece above it (Figure 23(b)). The lever and rods are held against the terminal pieces by the downward pull of the weight, but any relative movement between the weight and housing causes loss of contact.

This sensor too requires an analyser to control the current, set the sensitivity and sort out the true from the false alarms. It reacts if there are a series of small shocks, several medium shocks or one large one.

With other analysers the response can be pre-set to a given number of impulses in a specified time. Below this, the alarm is not triggered, so random shocks against a perimeter fence by children playing for example, have no effect. Some units have night/day changeover relays whereby the number of permitted shocks is reduced during the night when greater security is required and fewer random impulses are likely. Typical values are eight impulses in 30 seconds for daytime, and as low as four in 15 min at night. The values can be changed to give higher security (lower values) or greater immunity against false alarms (higher values) as circumstances dictate.

Inertia detectors are very sensitive, but in the event that a fence or structure is subject to continual or frequent small-scale vibrations such as from exposure to wind, magnetically damped sensors can be used. A similar effect may be obtained by reducing the sensitivity control. Frequency response is generally from 10 Hz to 1500 Hz and so covers the region generated by climbing or attack.

Inertia sensors are the best method of protecting fences, but they need to be placed at about 10 ft (3 m) intervals. This could mean quite a number if a large enclosed area has to be protected. However, a dozen or more can be connected to a single analyser and it is the analyser that is the most expensive item. Some analysers have separate zones which enable different sensitivities to be set for the respective groups of sensors. This is useful where sensors are used on different surfaces or positions. Also, indicators that latch-on, show which region produced the alarm.

Like the vibration detector they can also be used to protect large areas of glass, but their rejection of inoffensive vibrations make them more suitable for this task in active environments such as by main roads. However, even with these, the sensitivity setting is very much a case of trial and error. The best method is to set it for high sensitivity during the daytime, but with the analyser disconnected from the main alarm system. The latching indicator will probably soon show an alarm condition. Slightly reduce the sensitivity and leave for a further period, continuing the process until no alarm is shown at any time. Leave it at this setting for a few days to make sure a casual vibration or shock will not trigger it before connecting to the main system.

Inertia detectors can be used to protect walls and partitions and the spacing should be from 12 ft to 16 ft (4–5 m). Here they should be mounted at least 3 ft (1 m) away from floors, ceilings or abutting walls. For ceilings and roofs the spacing can be 20 ft (6 m) and the sensors should be mounted on beams and rafters. Some units are made for flush mounting into structural members while others are intended for surface mounting. In every case they must be positioned vertically and there is usually some mark on the casing to show the correct orientation.

Safe limpets

These are sensors that are designed to protect safes from attack, but could be used on filing cabinets or equipment consoles. Several types are obtainable, but most are fitted magnetically to the unit being protected (Figure 24).

The simplest type contains a vibration detector which operates if any attempt is made to force the safe open. It also has a contact which is held in place magnetically for as long as the unit is fixed to the safe. If it is removed, the contact opens and the alarm is triggered. Another type utilizes an inertia sensor which responds to low frequencies as well as the higher ones.

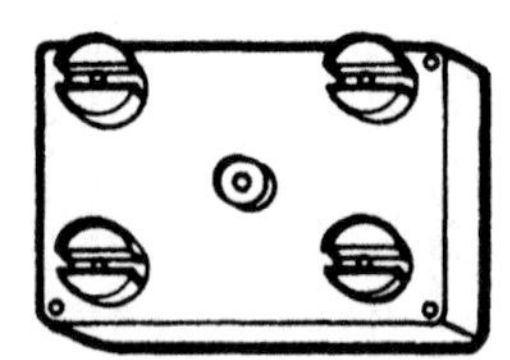

Figure 24 Safe limpet containing vibration and magnetic switch.

A third type includes a thermal switch as well as vibration contacts. This detects an abnormal rise in temperature as would be produced by a thermal lance or other metal cutting equipment.

Summary

It can be seen from the foregoing that there is a considerable armoury available in the form of various types of detectors. Several different ones, as we have seen, can be used for one specific application, while some are suitable for many. Each one has particular features that make it the best choice in a given situation.

All have one thing in common, they are perimeter detectors, that is they protect the outside boundaries of the premises and prevent intruders from getting inside without sounding the alarm.

It is always wise to have a second line of defence just in case the first one is somehow defeated. Also it is sometimes impossible to effectively protect the whole perimeter. The next chapter deals with further measures which are generally known as space protectors.

5 Space protection sensors

The detection devices, described in the previous chapter, were all (with the exception of pressure mats) operated by attempted penetration and movement of some part of the perimeter of the premises. For this reason they are often described as perimeter defences. It is essential, wherever possible, that the perimeter be fully protected, as it is far better to keep intruders out than to detect them once they are inside.

In some cases, full protection of all parts of the perimeter is impractical. Even where it is, where high security is needed, there is always the possibility that it could be breached by a determined and knowledgeable intruder. As a back-up, space protection detectors, often called *volumetric sensors*, can be added to the system to cover this need. They should not, though, be relied on as the sole means of defence. One reason, is the fairly high current of over 25 mA taken, which could not be sustained for long by ordinary batteries if the mains should fail. There are other reasons we shall see.

The main types of space protectors are ultrasonic, microwave, active infra-red and passive infra-red. There are also some lesser known ones that have special applications. The ultrasonic and microwave detectors rely on the Doppler effect, so we will firstly describe exactly what this is.

Doppler effect

The effect itself is quite familiar, the sirens of an approaching ambulance or police car sound higher in pitch than they actually are, yet when they pass and recede the pitch drops and sounds lower. This is due to the Doppler effect.

The pitch of a tone is governed by the number of sound pressure waves that reach the listener per second. This rate is termed the *frequency*, the unit of which is the *hertz* (abb: *Hz*; the multiple is the *kilohertz* abb: *kHz*, which is equal to 1000 Hz). When either the source or listener approaches the other, the waves come at a faster rate, so the frequency is higher. It is rather like swimming out to sea, you encounter more waves per minute than you would standing in the water waiting for them to come.

If the source and listener are receding, then we have an opposite effect, the rate is decreased, just as it would be if you swim or surf back to the shore; you may even encounter only one wave that carries you back in.

Pitch thus depends on relative movement between source and receiver. If an object reflects waves to a receiver, it in effect becomes a source, and if it moves toward or away from the receiver, a frequency difference will be perceived. This is the principle used to detect movement of an object in the protected space; waves from a source are reflected off any object that comes within range, and are picked up by the receiver.

Ultrasonic detectors

The term 'ultrasonic' denotes sound frequencies that are above the range of human hearing, the upper limit of which is 16 kHz in a healthy person in their twenties. The frequencies used in these detectors varies between different models, ranging from 23 kHz to 40 kHz. The frequency is generated by an electronic oscillator and is fed to one or more loudspeakers. These need to be very small, smaller in fact than the tweeter in a hi-fi loudspeaker, because the moving parts must move very rapidly and so their mass must be kept to a minimum. This is an advantage for an alarm system, as the units can then be made small and unobtrusive. The same unit contains a microphone with amplifying and processing circuits.

Ultra high-frequency sound is thus projected into the protected area. Some of it is received directly by the microphone from the loudspeaker, while some is picked up after being reflected from walls and objects in the room. When there is no movement, both direct and reflected sounds are of the same frequency. Should movement of an object occur, the sound reflected from it will undergo a change of frequency due to the Doppler effect. Thus the microphone now picks up two frequencies, the original received directly from the loudspeaker, and the shifted frequency reflected from the moving object.

If two different frequencies are mixed, the result is that a third one appears that is the difference between the two. When the two frequencies are close the third is often termed a *beat note*. The effect can sometimes be heard when two internal combustion engines are idling at slightly different speeds, the beat note is heard as a throbbing sound that varies in frequency as the speed of one or other of the engines changes.

The processing circuits detect the presence of this beat note and actuate the output relay which is of the normally-closed type. The device can thus be connected into a normal loop, but it is preferable for space detectors to be connected to a separate zone from the perimeter sensors so that rapid identification of the alarm source can be made. A power supply of from 25 mA to 50 mA is required from the control box.

Ultrasonic detectors usually have a range up to 27 ft (9 m), but they have a sensitivity control which can reduce the range down to about 10 ft (3 m). The polar pattern is in the form of a narrow lobe (Figure 25) and so the

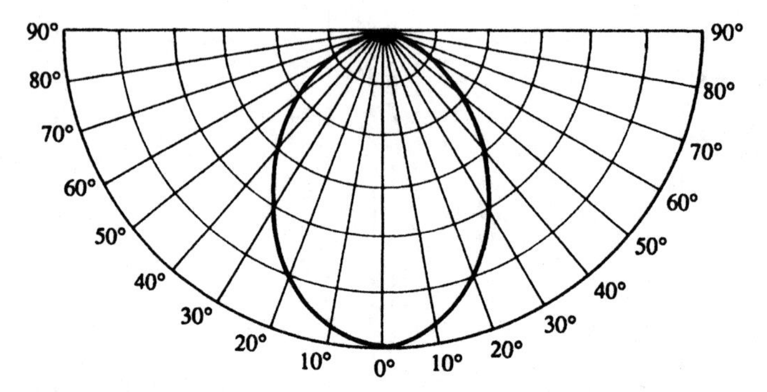

Figure 25 Polar diagram of the response of an ultrasonic detector.

device affords little protection at the sides. It should therefore be pointed toward the expected point of entry. As the Doppler effect is greatest for forward or backward motion, the sensor is more sensitive to these than to side-to-side movements. This is another reason to mount it facing the possible entry point.

Mounted in this position, the device is virtually impossible to defeat as any movement towards it, however slow and deliberate, will set it off. The only way to neutralize it is to do so in advance, but anti-tamper connections are usual as is also a microswitch which is released if the casing is removed. One possible way of tampering in advance would be to stick a felt pad over the front of the device. If of the same colour as the casing, this could pass unnoticed. To prevent this sort of action, the sensor should be mounted high and out of easy reach. Disguise is another defence; some units have been made to look like hi-fi speakers, office intercoms, and even a book.

The main disadvantage is susceptibility to false alarms. These can originate in several ways, ultrasonic sounds can be produced by vehicle brakes, gas or water jets, leaking compressed air lines, television sets and possibly computer visual display units (VDUs) and printers, among others. It is possible for almost any machinery to generate these frequencies as harmonics of their normal noise output. Such sounds will be picked up along with the sensor frequency and so produce a beat note which it will interpret as a Doppler product.

Another source of false alarms is air turbulence, which can distort the sound propagation and produce Doppler shifts. This will result from forced air heating systems, but can also be caused by other heating installations. Draughts, moving curtains and the like are other possible causes.

Modern ultrasonic detectors are somewhat less prone to these ills than the early models, as they have built-in circuitry for minimizing them. One type uses an averaging circuit which only responds to a net change in target distance. Vibrating objects, swaying curtains and some air turbulence average out to a zero position change, and so do not trigger the alarm.

The sensitivity control helps as this can be set to give only the necessary range. Reducing it in this way also reduces susceptibility to false alarms. However, even with such units, care should be taken in their use. It would be unwise to use them in an area where noisy night-time production was in progress nearby, in draughty premises, or those bordering a main road. They should never be used outdoors.

Microwave detectors

These use the Doppler effect and work in a similar way to the ultrasonic detector. The difference is that they employ radio waves that oscillate at extremely high frequencies. The standard is 10.7 GHz (1 GHz 1,000,000 kHz), but 1.48 GHz is also employed.

The receiver and transmitter are in the same unit, and the receiver receives some direct radiation from the transmitter, along with that reflected from our object within the protected area. If the reflected frequency is shifted by the Doppler effect, a beat frequency is produced and the result actuates the alarm circuits.

A feature of microwaves is that they will penetrate wood, glass, plaster, and to a limited extent brick. Microwave detectors also have a much greater range than the ultrasonic sensor, and will cover up to 150 ft (45 m). They are thus well suited for the protection of large warehouses, not only because of the size, but also because large stacks of cases and goods could afford shelter to an intruder from other types of volumetric detection. As microwaves pass through these, there is no shelter from them. Propagation patterns are usually tear-shaped in the horizontal field, with a narrow lobe in the vertical plane. Some units are designed with long narrow horizontal lobes to cover similar shaped areas, or have split beams to give V shaped coverage. Deflector plates can modify the pattern. (See the Technical Section, Figures 102–104.)

Steel cabinets, shelving or machinery could provide cover, because metals reflect microwaves. A solution to this problem is to mount the unit in the centre of the ceiling, as shown in Figure 26. Some models are designed for this purpose and have a hemi-spherical distribution pattern. The feature of these is that they can see over metal obstructions, so there is little cover for any intruder.

Coming from this angle the microwaves do not need to penetrate 'soft' materials such as wood or plastic, and so can be of a lower frequency such as 1.48 GHz. The range of a ceiling mounted unit at a height of 10 ft (3 m) is a circle at floor level of some 60 ft (18 m) diameter.

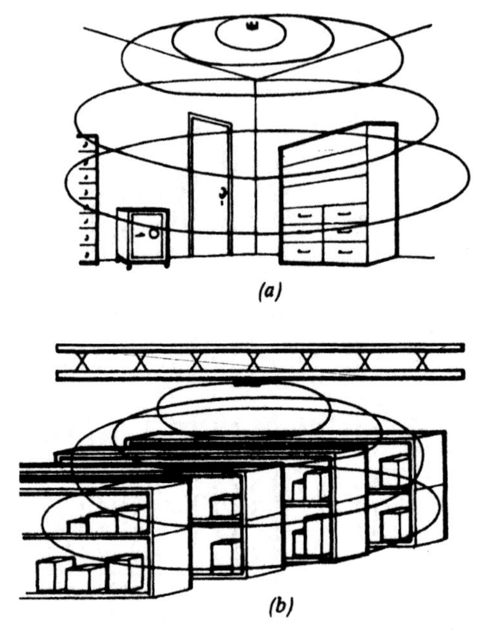

Figure 26 Ceiling-mounted microwave unit gives good coverage over limited floor area. There is virtually no penetration outside the area except through the floor. Ideal for ground floor areas. (a) Typical small office area protection. (b) Excellent coverage in factory storage area. Steel racking gives intruder shelter from horizontal beams, but none from overhead radiation.

Microwave sensors are not subject to the same susceptibility from false alarms as are ultrasonic detectors. Draughts, air disturbances or turbulence have no effect, nor have any type of sound waves. Some gas filled lamps such as fluorescent and neon lights can generate radio signals that could be accepted by the receiver, but there are usually filters that remove these electronically.

The main possibility of false alarms, comes from the penetration of microwaves beyond the perimeter walls, or in the case of the ceiling-mounted units, penetration of the floor to the room beneath. To reduce this possibility, the sensitivity must be adjusted so that the range extends only to the perimeter boundary. Even then it is possible for some overspill, especially through windows or doors. The effect can be avoided by curtailing the sensitivity, to range somewhere just short of the boundary, although this would reduce the level of security.

This really is the problem with most volumetric systems, and is the reason why they should be used as a back-up rather than the sole means of detection. If the perimeter is well protected, there need be no qualms in restricting the range to prevent false alarms.

Like the ultrasonic sensor, there is no way it can be defeated when it is switched on as any approach immediately triggers it. Tampering when switched off is the only hazard, but the normal anti-tamper arrangements should take care of that, except the possibility of a metal plate being attached to the front.

The units are usually totally enclosed in a plastic box without any apertures, as the microwaves travel through the plastic. So, it is not easy to tell just what the box contains, it could be merely a junction box. This anonymity is obviously a feature in its defence. Some ceiling models are flush fitting and so give no clue as to their true nature.

For applications where very high security is required, a microwave monitor can be installed. This can be installed at some position remote from the unit, but within its range. It detects the presence of microwaves, so if they are absent for any reason, it gives an alarm signal. This would also protect against inadvertent shielding by leaving a large metal object in the path of the microwaves, which could happen in a busy warehouse.

Microwave beam-breaking

Microwave Doppler systems are not suitable for outdoor sites as the perimeter is usually a wire or wooden fence which is quite transparent to microwaves, any moving object just beyond would activate the system. Reducing sensitivity to fall short of the perimeter would remove protection just where it was needed. Birds flying low across the site could also trigger false alarms.

Another type of microwave system does not use the Doppler effect, but transmits a narrow beam to a receiver situated at a remote point and thereby protects the space in between. The alarm is triggered if anything breaks the beam. The range can be considerable, up to 500 ft (150 m), and is very suitable for guarding outdoor sites.

Although a space protector, the space protected is a straight line and so the system is not volumetric. It is necessary for each boundary to have its own transmitter and receiver (Figure 27). This arrangement could guard up to 2000 ft (600 m) of boundary fencing and although four sets are required, this is far more economical than using inertia sensors every 10 ft (3 m). It has an inherent anti-tamper property, inasmuch as any interference with the transmitter prevents the beam from reaching the receiver.

The beam is not clearly and sharply defined like a focused light beam, but is rather diffuse. Its divergence can produce a beam width of up to 20 ft (6 m) at full range. Small objects passing through the beam are thereby defocused

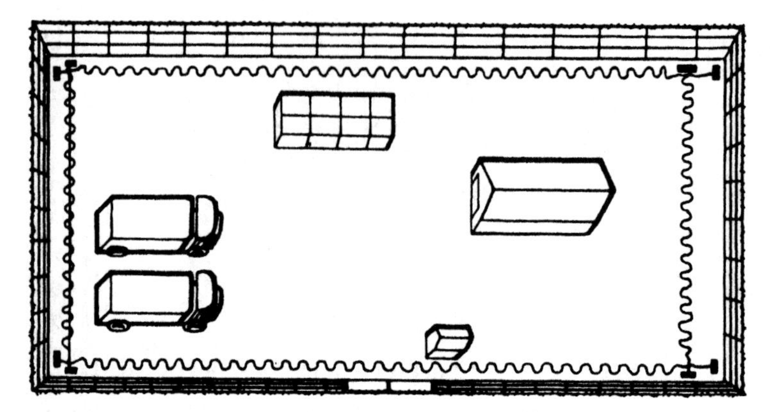

Figure 27 Perimeter fence with protection by four sets of beam-breaking microwave units.

and cast no shadow unless they are very close to the transmitter. The effect is similar to a fly in a headlamp beam. Thus birds and small windborne objects do not initiate an alarm condition.

The current required is rather high, 150–250 mA each detector, so to protect a site with four units takes about 1 A. This is more than usually delivered by control boxes and so a separate power unit is required. A standby battery would need to have at least the capacity of a car battery, which would last about 80 hours before recharging.

Active infra-red detectors

These operate in a similar way to the microwave devices in that a generator radiates a beam which is picked up by the receiver. If the beam is broken by an intruder the alarm is triggered. The infra-red rays are usually produced by a gallium arsenide crystal, hence they are sometimes called gallium arsenide rays, but they can be generated by a wide variety of means.

Any source of heat will usually produce infra-red radiation, because both lie just below visible light, in the same part of the spectrum. Thus, lamp bulbs, heaters and even a pocket torch generate them to a greater or lesser extent. This opens the possibility of defeating an infra-red system by the intruder directing a portable infra-red source at the receiver as he moved across and broke the original beam.

This type of interference is prevented by pulsing the beam. The frequency of the pulses vary, but 200 pulses a second is common. The receiver only responds to a pulsed beam of the same frequency, if the pulses are replaced by a steady beam, it signals an alarm.

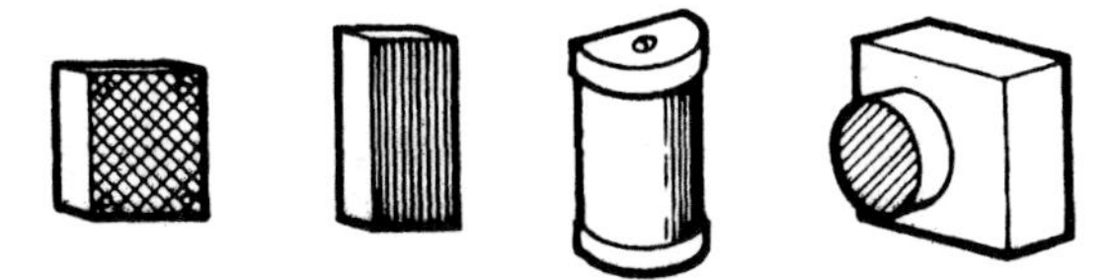

Figure 28 Selection of infra-red units. Projectors, receivers and mirrors are housed in identical cases. Dummy cases are also available to further confuse intruders.

The projector and receiver are housed in identical cases (Figure 28), hence it is not possible to identify which is which. This increases the difficulty for anyone trying to defeat the system, and the problem is compounded by the use of additional dummy housings which are available for the purpose.

The only way to evade the ray is by avoiding it, such as by crawling underneath it. To do this the intruder must be aware of its presence and its path, but even so evasion becomes impossible if the beam is laced across a vulnerable point in a zig-zag fashion. This can be done by means of mirrors; unlike the microwave beam which is too diffuse, the infra-red ray can be reflected at several points by small reflectors just like a ray of visible light (Figure 29).

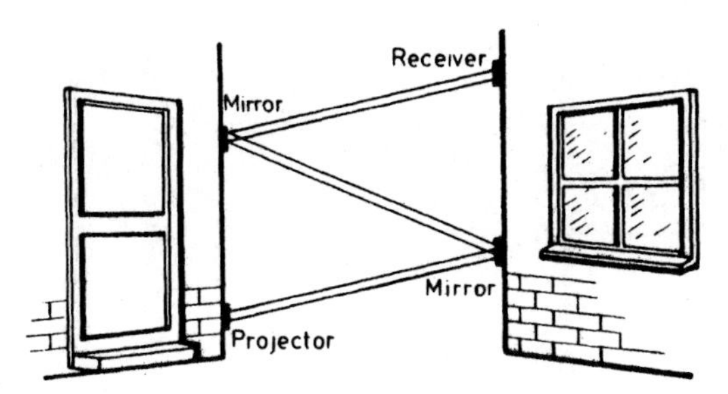

Figure 29 Lacing an infra-red beam across an area to make avoidance difficult. Lower units should be closely spaced to prevent intruders stepping through or crawling under the beams. False alarms are possible from wildlife, etc., to avoid this use two parallel beams.

The range varies from 65 ft (20 m) for the indoor type up to 1000 ft (300 m) for the most powerful outdoor type. When reflected from a mirror the ray is attenuated and so the range is shortened. For a single reflection the range is reduced by 25 per cent to 75 per cent; a second reflection reduces the range to 56 per cent, and third to 42 per cent. It is recommended that not more than three mirrors be used.

The rays will also penetrate clear glass and so protection can be extended through partitions and windows. Here again there is a reduction in range for each glass pane although the reduction is less than for mirror reflections. For 20 oz window glass the range is reduced by 16 per cent to 84 per cent for a single sheet; for a second sheet the range becomes 70 per cent and a third reduces it to 60 per cent while a fourth drops it to 50 per cent.

An infra-red beam is thus an alternative to the microwave beam for perimeter fence protection, and has the advantage that only a single projector and receiver need be used, with mirrors to divert the beam around the area instead of four microwave sets. Furthermore, the projector and receiver are in the same corner thereby saving wiring.

As the beam is narrower than that of the microwave, there is a possibility of false alarms due to birds or windborne objects interrupting it. This can be avoided by running two parallel beams, one above the other and connecting the receiver output contacts in parallel, so that both beams must be broken simultaneously to produce an alarm (Figure 30). This requires two sets, but there is a dual beam model that serves this function with a single unit.

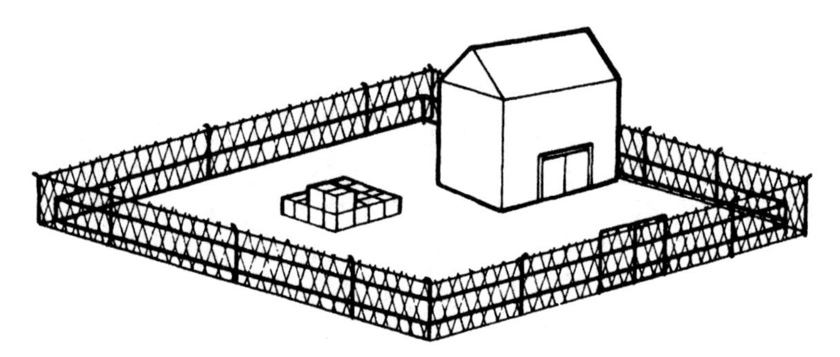

Figure 30 Infra-red perimeter system. Mirrors enable one continuous beam, but two beams are used in parallel to avoid false alarms. Should not be used in cold exposed areas where mirrors may be frozen over.

Another solution is the *double-knock analyser.* This is a circuit that can be wired to almost any sensor that could be subjected to false alarms. It triggers an alarm only after two alarm conditions have been received within a set time that can be adjusted from 10 s to 80 s. Security is slightly reduced by this but it greatly reduces the chance of false alarms.

A disadvantage of infra-red beams is that they are attenuated by fog and rain. Also condensation on the lenses of projector or receiver reduces the transmission through them. It is customary to include a heater in both units to keep them free from condensation, and this takes extra current from the supply. Heaters are not built in to units intended for internal use and this is the main difference between external and internal models.

The mirrors used to deflect the beam are passive and have no power supplies. Thus they have no heaters, and so can be affected by condensation or rain drops and ice and snow collecting on the surface. This is a serious drawback which could result in false alarms. The use of mirrors outdoors to achieve perimeter protection should be confined to locations where fog, ice and snow are rare, and even then the total distance should be well within the range of the unit to allow a generous safety margin.

An infra-red beam can be used to link an auxiliary alarm system in outbuildings to the main system as well as protect the intervening space. Figure 31 shows how this can be done. The sensors in the outbuilding are in series with a loop that carries the power supply for the infra-red projector. If they are actuated, the projector is switched off and the receiver on the main building signals an alarm to the main system. If the beam is broken, this also triggers the alarm. It should be noted that magnetic switches could not be used as the sensors, as these would not carry the current required to supply the infra-red projector. Microswitches would have to be used.

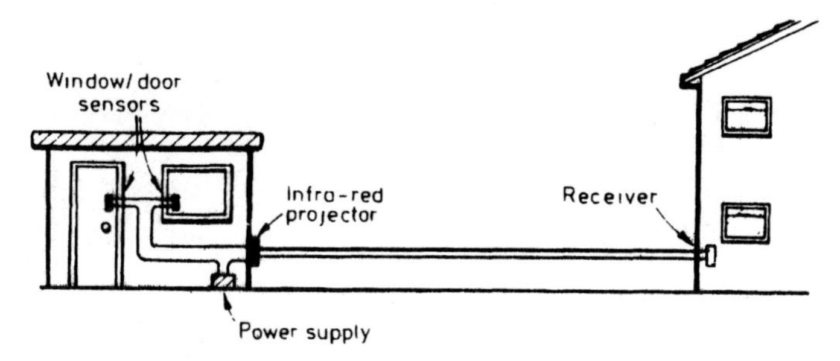

Figure 31 Separate alarm system in outbuilding can be tied in to the main alarm by means of infra-red beam. Sensor loop is in series with power supply to projector, so if sensors are actuated projector stops. Receiver on main building then sounds the alarm. Latching circuit is not necessary as this is included in main alarm, so alarm continues even if beam cessation is only momentary. Protection over intervening area is also achieved.

Indoors, the weather is not a problem, so mirrors can be used to deflect the beam around irregular areas (Figure 32). Infra-red beams can be used to protect stairways (Figure 33), and office or other work-station cubicles that are separated by clear glass panels (Figure 34).

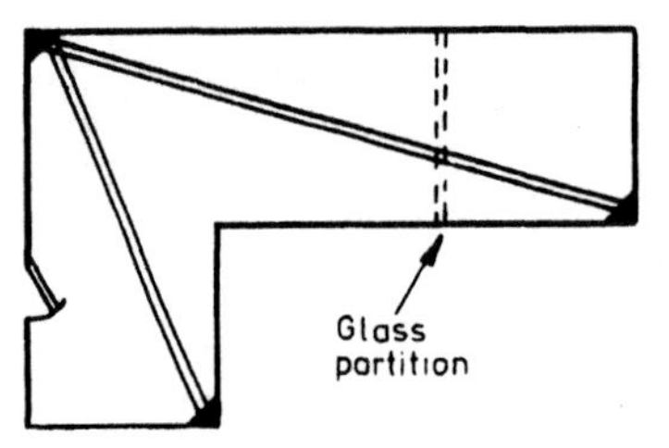

Figure 32 Infra-red beam with mirror used to protect irregular areas. Will pass through glass.

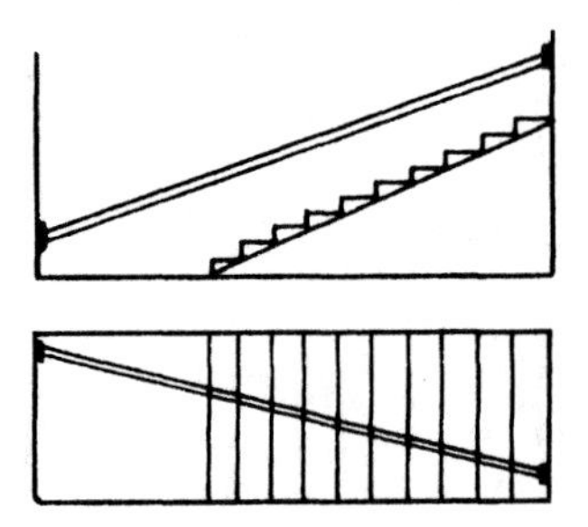

Figure 33 Stairway protection with infra-red beam. Diagonal angling prevents avoidance by keeping to one side.

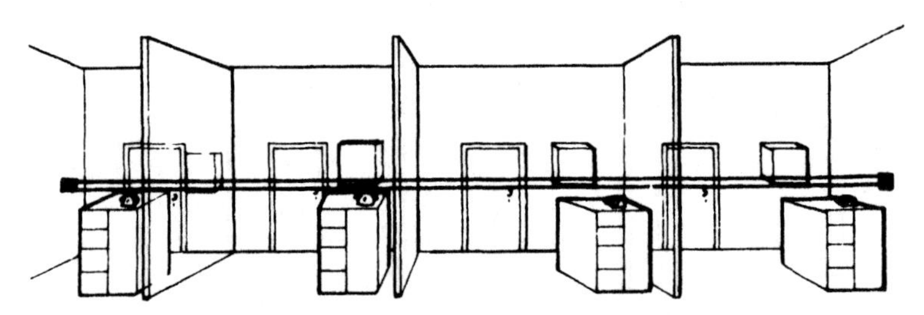

Figure 34 A divided office with glass partitions can be protected with a single beam. Opening any of the doors will break the beam. Check that the range is sufficient from Table 8 on page 277.

Passive infra-red detectors

The passive infra-red detector (PIR) is a newcomer to the security scene compared to the types we have so far discussed. It has been made possible by the development of highly sensitive ceramic infra-red detectors. The device does not have a projector, but operates by detecting the infra-red radiated by a human body.

The received radiation is focused on to the infra-red sensor, by either a curved mirror behind it, or a curved plastic lens placed in front. The curve is not continuous, but is broken up into a series of vertical facets which have a particular function. The detector has two sensitive areas and some facets focus on to one, while others focus on to the other. When a heat-emitting object moves across the field of view, the image thus appears alternately on these two areas. An electrical output is generated only when there is a varying difference between them, nothing happens when either, neither, or both receive radiation. Only a moving object thus produces an output, a stationary source of heat has no effect.

The facets produce detection zones like spreading fingers, and it is the crossing of these zones that produce the required fluctuations at the detector. There are usually two, or sometimes three, sets of horizontal divisions as well as vertical ones in the mirror or lens. These produce a second or third array of detection zones, one above the other and at different planes and angles. The effect is shown at Figure 35. These lower planes give protection to the space underneath the main zones and areas nearer to the sensor.

Detection is almost instantaneous and the range is surprisingly long considering the device relies on body heat. The usual range is 40 ft (12 m), but many have a range of 50 ft (15 m). Some special ones with a narrow field designed for corridors, have a range of up to 130 ft (40 m). The normal field

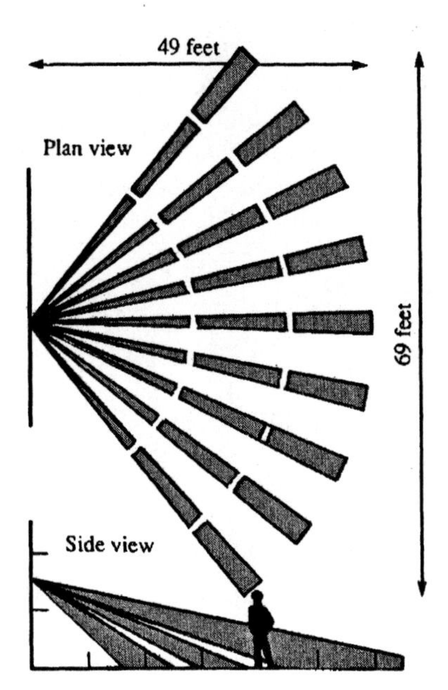

Figure 35 Typical polar pattern of PIR detector.

is fan-shaped with a 90° spread which is ideal for corner mounting, but there are variations, and some have a wider spread. (See the Technical Section, Figures 105–112).

When the unit is mounted vertically on a wall at a height of about 6–8 ft, the detection zones are angled downward so that the highest ones meet the ground at the rated range. A shorter range may be required such as for covering a yard bordered by railings. With small areas bounded by an open perimeter such as railings, the range could be too long, and false alarms could be generated by movement on the other side. A sensitivity control is usually included for range shortening, or more reliably, the device could be tilted downward, so that the detection zones are shortened.

It may be required to give greater protection closer to the unit in preference to that at a distance, and here again downward angling will achieve it by bringing the lower zones closer. Conversely, a slightly longer range than that specified may be obtained by mounting the unit higher, so that the upper zones meet the ground at a further distance. In this case the sensitivity control would have to be set to maximum.

False alarms due to inanimate moving objects are not possible, unless the object is warm, because the device is actuated only by objects that are both warm and moving. The detector is not affected by heating systems with the possible exception of forced air heating. In most cases this also has no effect, but it can under certain conditions.

While a strong infra-red ray from a projector will penetrate glass as we have already seen, that radiated by a human body is insufficient to activate a detector after passing through glass. So there is no danger of false alarms due to someone passing the other side of a window in a protected area.

In the case where random momentary triggering may be possible, models are available that have double-knock capabilities, they ignore a set number, which can be two to four disturbances, and trigger on the next. An intruder would certainly cross more than two zone boundaries, so there is little loss of security.

The most common causes of false alarms in outdoor applications are animals. Cats are no respecters of property rights and will, if entering the range of a PIR set it off. At least one model has only one plane of detection zones, and that is aimed horizontally with no downward tilt. It could be described as a moggy-proof model, but needs to be mounted no higher than waist-high to ensure detection of human intruders.

The response of a PIR to animals suggests a possible use in detecting foxes at poultry farms. For such an application the device should be mounted low but tilted back to achieve the range. Alternatively the moggy-proof unit could be mounted just above ground level. Human marauders would also be caught by it.

The PIR unit output is a relay with normally-closed contacts, that is they are closed when on guard but open when triggered or when the power supply is removed. Connection is thus made to a normal loop, and a power supply is required of from 12 mA to 25 mA. Anti-tamper connections are also provided thus necessitating a six-wire cable.

Resetting occurs as soon as the cause of the alarm has been removed, although of course the control box latches-on in the alarm condition. Some models have a latching facility. With these an extra wire back to the control panel is needed, requiring eight-core cable. An indicator light on the unit remains on, and the relay contacts stay open, after the cause of the alarm has been removed. By this means the unit initiating the alarm can be identified when two or more PIRs are used on the same control box zone.

PIR detectors are available not only as sensors linked to the main security system, but as auxiliary detectors, and most outdoor sensors are in this form. These are combined with floodlighting and an automatic daylight switch. The floodlights come on when a person or a vehicle comes into range, and the lightswitch ensures that the device is not activated in daylight.

With these, the device latches on and the lights remain lit for a period after the activating cause has ceased. The period can be pre-set from several seconds to many minutes, after which the unit resets and is ready for further triggering. If a body is still present and moving in the detection area, the light remains on.

Flood-lighting is a strong deterrent to intruders, but can be expensive to run all through the hours of darkness at a sufficient level of illumination. Lighting that comes on only when someone approaches is economical and also has an unnerving effect on an intending intruder. For legitimate callers and staff it is very convenient to have the outside of the premises illuminated on approach without the need for manual switching, and also serves as a deterrent against personal attack. Different parts of the outside area can be illuminated by separate units which switch on and off independently.

Another use of auxiliary PIR detectors is to give 24 hour protection to sensitive areas within the premises, against unauthorized intrusion. The detectors can be linked to a separate circuit to give internal warning to security or administrative staff.

It can be seen from this, that the PIR detector has many advantages over most other types of space protection. It has a high immunity from false alarms; it has a wider spread than ultrasonic or microwave detectors; it has a longer range than ultrasonic devices; it can be used outdoors; being based on optical principles, it can be angled to give exactly the coverage required, and accurately avoid adjacent unprotected areas. They are not, however, totally defeat-proof. They can be evaded by an intruder with a heat-shield made of aluminium foil or a photographer's silver parasol.

One characteristic of all PIR detectors is that they take a few minutes to stabilize after the power is switched on and so are not immediately on guard. Once stabilized, response to intrusion is instantaneous unless a double-knock node is operating.

Electrostatic devices

These are unusual and rarely encountered. The electrostatic sensor uses the same principle as occasionally seen in some shop window advertising displays. In these, a mobile display is connected to an electrode which is stuck on to the inside of the glass. Placing a finger over the electrode on the outside, introduces a capacitance to earth because the human body is virtually an earthed mass. This capacitance affects a delicately balanced electronic circuit which trips a relay and switches on the model.

In the case of the alarm, an electrostatic field is generated around the sensor. Any mass introduced within the field upsets the balance, and the associated electronic circuit triggers the alarm. The range of the device is limited to a few feet, which, when compared to other systems is a disadvantage. However, it can only be actuated by a mass of predetermined size, so cannot be triggered by any of the usual causes of false alarms. It could be used as a back-up to protect limited areas such as in front of safes, valuable paintings or trophy cases.

Summary

For indoor space protection the choice lies between ultrasonic, microwave, active infra-red and passive infra-red. The electrostatic system is a rarity and can be ignored other than for a third line of defence behind one of the others where very high security is necessary.

Ultrasonic detectors were once the main choice for space protection but have fallen from favour since the appearance of other types, especially the passive infra-red. Their main disadvantage is their susceptibility to sources of false alarms. They can be triggered by noises having harmonics in the ultrasonic range. Also they can be activated by draughts, by innocuous moving objects such as curtains, and by air movement generated by heating systems. While some units have a degree of immunity to these by only responding to net movement, there is still a higher risk of false alarms.

Another drawback is their range which at around 27 ft (9 m) is less than microwave or PIR detectors, furthermore the sideways coverage is restricted to a narrow lobe.

Like the microwave detector, the use of the Doppler principle gives the device a high sensitivity to movement toward or away from the detector. It is less sensitive to sideways movement. The PIR sensor has the opposite characteristic, being more sensitive to sideways movement. Thus an ultrasonic detector gives good protection in corridors and other long narrow areas where the intruder would be approaching the unit.

As it responds to movement by any object the ultrasonic detector is virtually impossible to defeat. It is theoretically conceivable that a PIR detector could be evaded by an intruder carrying a large object in front of him as a heat shield, although whether it would be practical is a different matter. So, providing no possible sources of false alarms are present, the ultrasonic detector can be used to give a high degree of security.

Microwave systems are not susceptible to the interference sources that afflict ultrasonic units, but they have others. The main one is penetration of boundaries especially through doors or windows. They can thus be triggered by movement outside the perimeter unless the sensitivity is carefully adjusted.

Solid-object penetration can be an advantage in places such as warehouses where there could be many large objects to provide an intruder cover from other types of detector. In this situation, ceiling-mounting is the best, and this also eliminates boundary penetration other than through the floor. The range obtainable from microwave detectors is greater than that offered by ultrasonic devices.

Active infra-red systems, although included in this section, are not volumetric protection devices as they guard only a narrow area through which the projected beam travels. Most of the possible applications are better served by PIR detectors, but there are some for which the beam system could be preferred.

Unlike the PIR sensor it is effective through several layers of glass, and so a single device could be used to protect several glass-walled cubicles. Furthermore it can be used to protect items at exhibitions, boundaries beyond which the public can approach to view, but not exceed, or anywhere where invisible boundary protection is required.

The PIR detector has taken the place of most of the other types for the majority of applications although there are a few such as those mentioned above, for which others are better. Responding only to objects that both move and generate infra-red radiation eliminates many causes of false alarms in other sensors. A wide, normally 90° angle, and a range of up to 50 ft (15 m), gives protection over a much larger area than most. It can be precisely aimed to give just the coverage required and so not encroach on outside areas.

Although most sensitive to sideways motion, it is by no means insensitive to forward and backward movement. As the detection zones spread out fanwise from the unit, and most detectors have two or three vertical planes of detection, it is virtually impossible to move more than a few inches within range without traversing across two or more zones.

While the possibility of avoiding detection by using a heat-shield has been suggested, it is a remote one. It would mean that the intruder would have to know that PIR detectors were installed and where, he would need a knowledge of how they work, and come equipped with a shield that was large enough, light, and manoeuvrable. While this is quite possible for the professional thief after a high prize, it is very unlikely for the casual opportunist burglar. Even so, where very high security is required, another system such as a microwave or ultrasonic one should be used as well as the PIRs.

There are some dual technology units that combine a PIR with either a pair of microwave detectors, or an ultrasonic detector. Some of these can be set either for *both* sensors to be actuated in order to trigger an alarm, or for *either* of them. In the first mode, any condition which may produce a false alarm in the one may not affect the other, so conferring a high immunity against false alarms which is useful in difficult environments. In the second mode, an intrusion missed by one, could be detected by the other, thereby increasing the security. The choice can thus be made according to circumstance. One unit combines a PIR with an acoustic breaking-glass detector, but in this case they act independently.

For outdoor applications the choice is narrowed to microwave, active infra-red and passive infra-red. To protect a large perimeter fence, four microwave beam detectors are required. A single infra-red beam could do the same job using deflection mirrors, but there are snags. First, although having a range of up to 1000 ft (300 m) this drops considerably if reflected three times. The range is reduced to 400 ft (120 m). So, this gives a rule of thumb, that to enclose four sides of a perimeter with a single beam reduces the range to 40 per cent of the original.

However, to avoid false alarms due to small objects breaking the beam either a twin-beam system or a double-knock accessory is required. The latter is more economical, although it means a slight reduction in security. Another problem is the effects of weather, particularly condensation, ice and snow. Built-in heaters keep the projector and receiver clear, but not the reflectors. Microwaves are not affected by these, and are thus the best method of protection where adverse weather conditions are likely.

Passive infra-red detectors can be used for outdoor protection, but the problem is that of stray animals. Sensors having non-tilted detection zones by directing them parallel to the ground considerably reduce the possibility of false alarms from this cause, providing the animal does not jump up into the plane of detection. A double-knock or pulse-counting facility should further reduce the risk. The floodlight type have much to commend them here as false alarms are of little consequence. Even if PIR detectors connected to the alarm system are used outside, it is well worthwhile augmenting them with PIR floodlights.

The range of a single detector may be insufficient to cover a large outside yard, but a combination, such as one in each corner, could protect up to 100 ft (30 m) long sides of a square and most of the internal area (Figure 36).

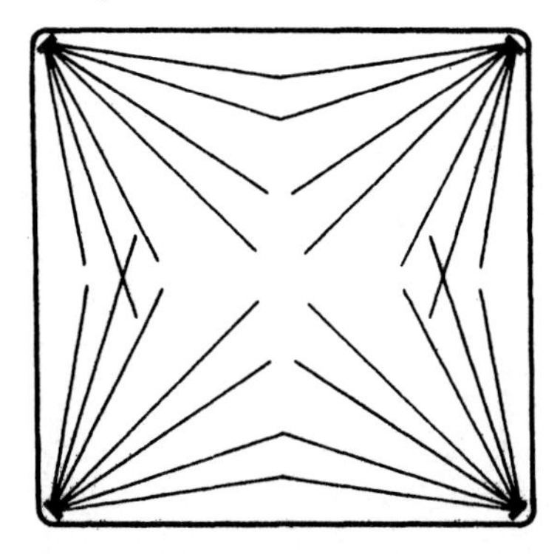

Figure 36 A large area can be covered with four PIRs in the corners.

For distances longer than this, beam, microwave or active infra-red systems should be employed. For perimeter fence protection though, inertia sensors are the most effective.

Table 1 Table of motion detector sensor characteristics

Condition	*Ultrasonic*	*Microwave*	*PIR*	*Infra-red beam*
Draughts, turbulence	×	✓	†	✓
High pitched sounds	×	✓	✓	✓
Heaters	×	✓	†	✓
Moving curtains, etc.	*	†	✓	✓
High humidity	†	✓	†	†
High temperature	†	✓	*	✓
Static reflections	✓	*	✓	✓
Sunlight	✓	✓	†	†
Movements beyond perimeter	✓	*	✓	✓
Vibration	*	*	✓	✓
Water in plastic pipes	✓	*	✓	✓
Small animals	*	*	×	†
Mutual interference	†	†	✓	✓
Long range	–	✓	–	✓
Wide coverage area	–	✓	✓	–

Key: ✓ No problems; † Slight problems under certain circumstances; * More serious problems, needs careful setting up and sitting; × Major problems, not recommended with this condition.

6 Sounding the alarm

It is a thought provoking fact that all the technology, thought, planning and work that has gone into the design and installation of a security system, with its sensors and sophisticated control circuitry, is in most cases all to the end of just ringing a bell. Of course, it is the effect that the bell has that is the important thing, but it does emphasize the importance of the sounding device. If it is ineffective or can be neutralized, the whole of the rest of the system is useless.

As stated earlier, it is prudent to have at least two bells if not three. These can be situated at the front and back of the building and also inside. Even if the outside bells were ignored or incapacitated, few intruders would have the nerve to enter premises in which a loud alarm was sounding.

The important thing with any alarm sounding device is that it should be loud and also strident. Not all bells score well on both points. Some have quite a modest sound output compared to others, while many sound quite melodious and not at all nerve jangling as they should.

Sound levels and loudness

To understand the volume ratings of different sounders we need to know something of how sound is measured. The decibel (dB) is the unit used, but it does not express an absolute value such as an inch or a pint. The reason for this lies in the way our ears perceive sound. Loudness follows a logarithmic rather than a linear scale. So a sound that measures twice the level of another, does not seem twice as loud to our ears, and one that sounds twice as loud would actually be many times the level of the other.

So to specify the aural effect of a sound in loudness, we have to use units that compare the measured sound to a standard reference level, and in addition progress in a logarithmic manner. The reference level is the human hearing threshold in a healthy young adult, which is a sound pressure of 20 μPa. All other sound levels are thus expressed as a ratio to this which is thus given the value of 0 dB.

Anything over 80 dB is usually considered loud, while 100 dB is unpleasantly so. The threshold of pain is reached at 130 dB and exposure to this level can cause permanent hearing damage in a matter of minutes.

The decibel can be used to specify a level difference between two sound sources such as two bells, or the difference in level at different distances of the same source.

To get some idea of the levels of everyday sounds the following values are given:

Inside room in quiet neighbourhood	20 dB
A watch ticking	30 dB
Spoken whisper at 1 m	45 dB
Good speaking voice at 1 m	60 dB
Vacuum cleaner at 1 m	75 dB
Average disco level	100 dB
Pneumatic drill at 1 m	105 dB

Some common decibel ratios are:

6 dB × 2
10 dB × 3
12 dB × 4
18 dB × 8
20 dB × 10

Sound levels decrease with distance and this is important when considering the siting of a bell and also comparing specifications. The levels of some sounders are specified at 3 m, while others are specified at 1 m. This makes them appear louder than they really are when compared with those using the 3 m specification, but it is the one now most commonly used.

Sound pressure for most sources drops in proportion to the distance, so at twice the distance the pressure is a half and at three times the distance it is a third, and so on. The dB value for a ratio of 1 : 3 is, according to the above table, 10 dB. So a bell rated at 90 dB at 1 m is the same as one specified at 80 dB for 3 m. This point should be carefully noted when comparing specifications.

While the bell needs to be mounted high enough to be out of reach of tamperers, it should not be so high that the sound level is excessively attenuated. At a height of 6 m it is only a sixth of its quoted rating at 1 m, or 16 dB less. This underlines the point made about having a sounder at the rear as well as the front of the premises. The distance from the front to the back along the shortest path could be 20 m or more, and the sound thereby drop to a twentieth, or 26 dB less.

While the sound may be rather more than these figures suggest, due to reflections from the walls of the building, if the premises are detached and remote from others or from a frequently used thoroughfare, the bell may not be heard at all by anyone who could summon help. Intruders may take a chance on that and ignore the sounding bell, hoping to be in and out before someone eventually hears it. In such a situation, other warning devices are available which will be described later.

The minimum sound level at which a sounder can be deemed audible above moderate ambient noise is 60 dB. Table 9 in the Technical Section (p. 282) shows the theoretical distance at which 60 dB will be obtained for a given level at 3 ft (1 m).

This is in still air, any breeze will make a considerable difference, increasing the range in the direction to which it is blowing and decreasing it in all others. The normal temperature gradient whereby air gets cooler with height, causes sound to refract upwards and so is reduced at ground level. This also reduces the range. Sometimes there is a temperature inversion when air above the ground is warmer than at ground level. This causes downward bending of the sound waves and increases the range at which sounds can be heard. The table is thus a rough guide, as much depends on local atmospheric conditions.

Bells

The most common alarm sounding device is the bell. The operating principle is quite simple, a current applied to a coil, termed a solenoid, and a plunger is magnetically attracted through its centre to strike the rim of a gong (see Figure 37). As it does so a pair of contacts are separated and the current through the coil interrupted. The plunger is returned to its rest position by means of a spring, and the contacts close thereby reapplying the current so causing a further strike.

The distance travelled by the plunger determines the quality of the ring, if it is too close to the gong, one stroke is dampened by the striker returning

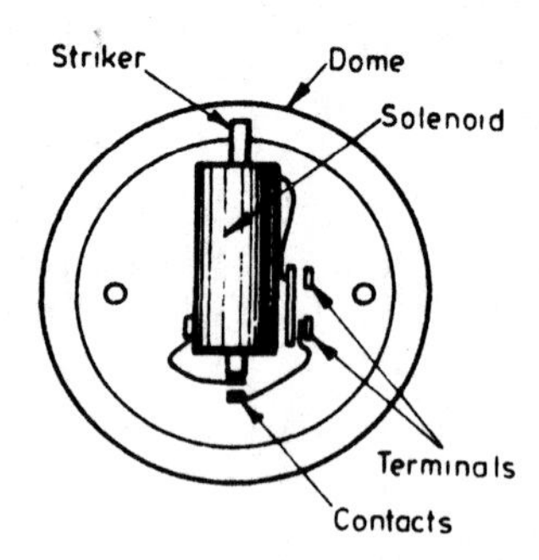

Figure 37 Basic principle of underdome bell. When solenoid is energized, striker is drawn upwards hitting the dome. Contacts are also parted, thus de-energizing the solenoid and allowing the striker to return to original position for the next stroke.

too soon for the next, resulting in a clattering sound rather than a ring. If the plunger is too far away, it does not hit the dome with the full force and the strokes come at too slow a rate, so producing a weak sound.

There is thus an optimum distance which can be set by a screw adjustment of the contacts or rotating the slightly eccentric gong. The older type of bell had the striker mounted on an arm with the coil and contacts located at the side of the gong. Modern bells, especially those for alarm work, have everything concealed underneath it.

This type of bell, with interruptor contacts, will work off a.c. or d.c., because polarity does not affect the action. They can thus be powered from a mains transformer of appropriate voltage output if a simple warning system is required that is operated by a push switch, or any sensor other than a magnetic switch. Bells intended for a.c. operation do not have interrupter contacts, and some are designed to work directly from the mains supply, although these are generally not used with intruder alarm systems.

The control box supply is usually 12 V d.c. Other bell ratings are 6 V, 24 V, and 48 V, the higher ones being mostly used for fire detection systems. These voltages are not critical and many bells are rated over a range such as 6–12 V, 12–24 V. With these, two current and sound pressure ratings are given.

The sound output is greater at the higher voltage within the stated range, as is the current. Current rating is important where several bells are to be used on the same system as the total must not exceed that specified for the control box. Also important is whether the standby battery could sustain an alarm with all bells sounding for the full alarm period (usually 20 minutes), and then be ready for a further alarm without replacement or recharging.

A bell should thus be chosen that has a moderate current rating but high sound output. A good bell should take less than 100 mA (0.1 A) at 12 V, and deliver in the region of 96 dB at 1 m. Not all models achieve this standard, either taking more current, delivering less sound, or in some cases both.

The design of the gong has much to do with the sound level and also the tone. Those having a dished effect in the centre appear to give a harsher and louder sound than the conventional dome shape, as do also those made of a heavier gauge steel. Most alarm bells have a gong that is 6 in (152 mm) in diameter, but some are also available with 8 in (203 mm) and more rarely, 10 in (254 mm) gongs. An 8 in of the same make usually has the same coil and striking system as the smaller version but gives about a 3 dB greater sound volume, which is over 40 per cent more. This does not necessarily follow when comparing different makes, as the larger gong may be of thinner gauge steel, hence of smaller mass, and could thus actually produce less sound than the smaller one.

Another type known as the *centrifugal* bell, uses a small motor to rotate the striker at high speed, repeatedly striking the inside rim of the gong. Current is usually higher than for the best solenoid types, although it is about the same as most of the high-current ones. Sound output is about the same. The construction is less robust than the solenoid bell with more moving parts, so there seems little to recommend it over the solenoid type.

Enclosures

It is the common practice to enclose the bell in a case. Materials used are polypropylene which has moderate strength but does not rust, polycarbonate, which is also rustproof but much stronger, plastic coated steel for maximum strength, or stainless steel. Enclosing has two objects, to protect the bell from the weather and also from tampering (Figure 38). To increase security, cases are usually fitted with anti-tamper microswitches under the cover.

Figure 38 (left) Steel bell box with side louvres, (right) totally enclosed fibre-glass bell box.

Inevitably, enclosing the bell reduces the radiated sound. This is generally considered to be the inevitable price to pay for the security and weatherproofing a box affords. However, vented boxes can actually be a security hazard as a bell can quickly be stifled by sealing foam squirted through the vents. Boxes are not as indispensable as they are often thought, a far greater protection is to mount the bell in a position that is inaccessible without a ladder.

If access is obtained, many bells could indeed be easily neutralized by removing the gong, but there are models that have a centre gong-screw that can only be removed with a special tool, and are virtually impossible to silence by jamming or physical attack (Figure 39). One maker of such a bell is Tann Synchronome. These are also weatherproof and are well worth considering instead of the more usual enclosed bell. A further advantage is that it is evident that an alarm system is indeed installed. Dummy bell boxes are frequently fitted as a deterrent, but the more knowledgeable of the criminal fraternity are getting wise to this bluff. An open bell leaves no doubt.

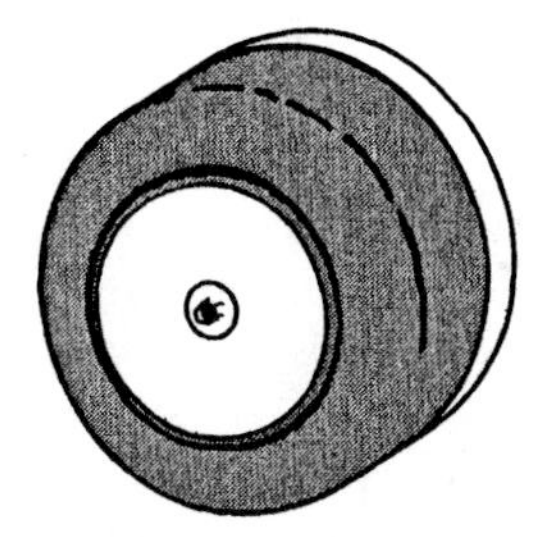

Figure 39 Tann Synchronome bell.

Self-activating bells

The wiring to the bell may be vulnerable to tampering in some installations. It can be protected to some degree by means of anti-tamper loop wiring as are the sensors, even if an enclosure with an anti-tamper microswitch is not used. However, as the bell has such a vital role to play, special measures are needed to combat possible attacks on the bell wiring.

One method of doing this is to use a self-activated bell. The bell is housed in a box with its own battery and a hold-off circuit. The circuit is effectively held off by a voltage from the control unit, so if the wire is broken or short-circuited, the hold is released and the bell is sounded.

Actually, more than just the wiring is protected by modern units. If the control box is destroyed, or the power supply cut-off, or the system standby battery runs down, it has the same effect, a releasing of the hold-off and sounding of the bell. It thus affords a back-up to the whole alarm system.

At one time, dry batteries were housed in the bell box and were the weak point of the system. Unused batteries deteriorated and needed replacement,

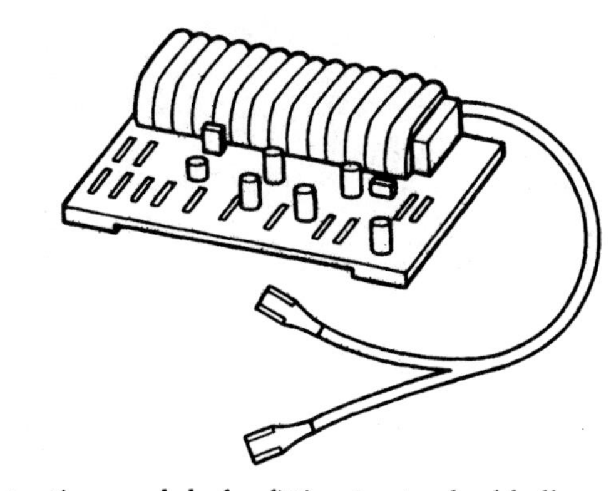

Figure 40 Self-actuating module for fitting to standard bell.

but were not easy to test or replace. With modern units, rechargeable NiCad batteries are used, and these are automatically kept charged from the control unit. A further feature is that a normal alarm, that is, one not involving operation of the self-activating circuitry, does not use current from the battery, but draws it from the control box in the usual way. The battery is thus kept fully charged for any occasion when it may be actually needed.

Self-activating modules are available which can be used in conjunction with most bells (Figure 40). Most, though not all, come complete with the battery fitted to the board. The modules are usually housed in the box with the bell, and an anti-tamper microswitch is usually included (Figure 41). It is possible to use a module with an open bell, by housing it in a locked case on the inner side of the wall on which the bell is mounted. Wiring from the module then passes straight through the wall to the bell.

Figure 41 Self-latching bell in box. Latching module and terminal strip can be seen, space is for housing internal batteries.

When self-activated, the bell is stopped by reapplying the hold-off voltage. If this ceased because of damage to the wiring or the power supply it may take some time to restore. However, the on-board battery capacity is limited to less than an hour of sounding, depending on the bell current, so there is no fear of the bell continuing to ring for hours until someone can deal with it.

Although security is increased by self-activation, it is not totally defeat-proof. As hold-off is obtained by a voltage supplied from the control unit, it can also be achieved by connecting an external battery across the hold-off circuit. This can be done by means of a battery connected to the appropriate wires via pins pushed through the insulation. The alarm circuit can thus be neutralized while the wires back to the control box are cut. If left so connected, the battery can hold off the alarm for several hours.

To succeed, the correct pair of wires out of the four usually run to the bell would have to be intercepted, and the polarity of the battery would have to be correct. This could be discovered by pinning the four wires and checking with a voltmeter to determine which pair carried the hold-off voltage, and its polarity. Not the sort of thing the casual amateur thief would be likely to do, but not past the ingenuity of a professional with a high prize in view.

The best protection for bell wiring is to conceal it where it cannot be reached, in conduit, under flooring, or buried in plaster; it should always be brought out directly behind the bell and never run along part of the wall exterior. If this is done, self-activation may be deemed unnecessary for most security systems. Where extra protection is needed, self-activation can be used, but the bell wiring should also be physically well protected and hidden.

Timing modules

As pointed out in an earlier chapter, it is desirable and in many places mandatory, to limit the bell sounding to around 20 minutes. The intruders should have disappeared by then, either under their own steam or in the back of a police car. Further sounding only causes annoyance to those nearby, especially at night, and if, as it often is, due to a false alarm.

All modern control panels have integral timers which so limit the bell sounding, but many older ones do not. If such an untimed system is already installed and is satisfactory in all other respects, there is little point in changing it. A timing module can in most cases be fitted to exercise the necessary control. Timing is usually adjustable up to around 45 minutes maximum. In some more remote sites such as on trading estates away from residential areas, a longer sounding than 20 minutes may be desirable to ensure that attention is attracted.

Sirens

Where extra loud warnings are required or they must sound over much longer distances than usual, a siren is often used. The mechanical type consists of an electric motor driving an impeller which forces air through vents in the casing in such a way as to produce a loud tone, usually of a raucous nature to arrest attention. The complete unit is totally enclosed and mostly is designed for outdoor applications (Figure 42), although smaller non-weatherproof versions are available for indoor use.

Decibel ratings usually range from 100 dB to 120 dB at 1 m which is louder than a pneumatic drill. Makers commonly quote distances at which the device is audible. These usually assume a minimum audibility requirement of 60 dB and so generally correspond with Table 9 in the Technical Section. As already mentioned, these distances depend on local conditions. It will be noticed from the table that there is a considerable difference in distance for just a few dB, especially at the high end. For remote locations then, a few dB extra can make all the difference as to whether the alarm is heard or not.

The high-power models are usually driven by mains voltage, and so most control panels require a relay to run them. Some panels have a switched

Figure 42 (left) High power horizontal siren, (right) smaller vertically mounted siren.

mains outlet to which these sirens can be directly connected. It should be noted though, that should the mains fail, and the system switch itself over to the standby battery, no alarm can be sounded if the sole sounder is mains operated. A bell or 12 V siren should also be installed to sound in the event of an alarm during a mains failure.

Sirens with a 12 V motor can operate directly from most control panels which deliver 1 A to the sounder, but some caution is needed. Average siren current is usually just under 1 A, often 0.9 A which is within the control panel rating, but the current at the instant of starting can be a couple of amps or more. It is as well to check with the control panel makers if the panel will take this momentary excess load.

If such a siren is used, no other sounder can be connected as the total output current available is thereby taken. If others are required, a relay will be necessary that will switch the total load to be applied.

An alternative is to use a solid-state electronic siren (Figure 43). These consist of an oscillator, amplifier and horn-type loudspeaker all contained in the same unit. Output is slightly less than for the loudest motorized sirens, but the current is much less. Some models take as little as 20 mA (0.02 A) yet produce up to 107 dB at 3 ft (1 m). Others which deliver some 116 dB, take 350 mA (0.35 A). Although the latter are comparable in sound output to the motorized sirens, the current taken is little more than a third. They have no starting current problems, and other additional sounders can be comfortably accommodated within the 1 A maximum output of the control panel.

Figure 43 Electronic sirens.

A horn speaker is used because this produces more volume per watt than any other type. To take the minimum space, the horn is re-entrant, that is folded within itself. However, it is a highly directional device and so needs to be pointed toward the area from where attention is to be attracted.

The frequency, or pitch, of the generated note varies between models. The ear is most sensitive to frequencies between 2–4 kHz, however, tones in this region and above are attenuated with increasing severity as distances extend beyond 165 ft (50 m). So at long distances the losses cancel any advantage gained by using frequencies in this band. Those just below it, from about 800 Hz to 1 kHz, are still very effective, yet they carry considerably further. They are therefore the most commonly used.

Electronic sirens often offer a choice of effects. The most common are: a fixed steady tone; a warbling effect between two tones; or a single tone that is pulsed about twice a second. These can be selected at will by using the appropriate connections, and they offer the possibility of having different sounds to indicate different things. One could be used to warn of intruders, another for fire, and in the case of a factory, a third (perhaps the steady note) could be used as a finishing signal.

Another type of electronic sounder is the piezo electric siren, which uses a piezo electric transducer instead of the horn. These also give a high output with low current consumption. Their main disadvantage is that the tone is high-pitched, which although it attracts attention, does not carry very well outdoors. They can be used effectively for short-range outdoor use, or as an internal sounder.

The electronic siren is thus versatile and easily matched to any control panel. It is of particular value in industrial situations when large areas such as factories, goods yards and large warehouses need to be covered, or buildings which are remote from any habitation. For High Street commercial applications such as shop and office premises though, the bell is to be preferred. In city streets, a siren can be confused with vehicle alarms, police car, ambulance and fire tender sirens. When an insistent bell is heard, there is little doubt as to what it is, an intruder alarm on nearby premises.

Illuminators

Although not strictly *sounding* the alarm, some form of illumination or visual indication linked to the alarm system is often desirable. Consider a common situation where there is a row of shops each having an alarm bell outside. Suddenly one starts to sound, but which one? It is frequently difficult to determine, and often requires the listener to stand under each one in turn to identify it. A visual indication will quickly show which premises are being attacked and so help can be summoned with minimum delay.

This indication can be arranged by means of a lamp mounted on or near the bell box. Some boxes are translucent so the lamp can be mounted inside the box if there is room. The light source commonly used is the xenon tube, a tube filled with xenon gas which emits a brilliant flash when a high voltage is momentarily discharged through it. The high voltage is produced by the integral associated circuit in the beacon, but the actual supply voltage is usually 12 V, although there are also 24 V and mains voltage versions. The light is commonly, though incorrectly, called a strobe.

The brightness is determined by the electrical energy discharged through the device and is rated in joules. Typical value is 5 J, but there are higher powered ones up to 12 J. The joule is a power of 1 W sustained for 1 second, but as the flashes last for only a fraction of a second, the light intensity is equivalent to that of a high-wattage filament lamp. Even the 5 J devices can be visible for several miles, depending on background ambient illumination.

Normally, the flash rate is about once per second, but it can be set to a higher rate of up to two a second if required. The life of the xenon tube is some 5 million discharges or more, so at the highest rate of 120 a minute, and for a period of 20 minutes operation for each alarm, it should last for at least 2000 alarms. It is hoped that no one would be so unfortunate as to ever need a new tube!

The current taken by the 5 J lamp is in the region of 100 mA (0.1 A), and so the unit can in most cases be connected directly across the bell without exceeding the control panel output current. The larger ones take 2 A or more at 12 V and so require an output relay to operate from a normal system. If the control panel has a mains switching facility, a mains-voltage xenon could be used directly if high power is required, without the need of an extra relay.

As well as giving visual warning and identifying the source of an alarm, lamps can be used to provide steady illumination or to terrify and confuse the intruders. The control panel mains switching facility or a separate relay can be connected to high power filament lamps to floodlight a yard or the area around the premises. Several lamps can be used providing the total wattage does not exceed that specified for the relay or panel. These will remain on for as long as the alarm is activated.

Another interesting device is the flasher module, which is connected to the alarm output circuit. This consists of a relay that operates a changeover switch once per second. Two high-power (up to 500 W) lamps can be connected so that they go off-and-on alternately, one comes on when the other goes off. If the lamps are situated well apart, the alternate illumination from different angles gives the impression of being surrounded by flashing light and will wreck the nerves of the boldest intruder, especially as the alarm will be sounding at the same time. All these devices are linked to the alarm system and so come on when the alarm is triggered.

The passive infra-red controlled lamps described in Chapter 5 which operate independently, are also worth considering. These are activated by body heat over a range of many feet, and switch on the associated floodlight whenever they detect movement of a warm body within their range. They can thus be used within the protected perimeter as a back-up system, or at the perimeter to discourage attempts at entry. Once activated, the light remains on for a pre-set period of from seconds to minutes.

A further type of security illuminator is the rotating mirror beacon such as used on police cars and ambulances. These are available in different colours, though not usually blue to prevent misrepresentation. The lamp is rated at 48 W, but the parabolic reflecting mirror concentrates the light into a powerful rotating beam. This is most effective when there is space all around rather than when the beacon is fitted to a wall.

Remote signalling

Our concern, so far, has been for the alarm system to actuate an on-the-spot sounder. Its purpose is to scare off the intruders before goods can be removed or damage committed, and to summon help. The help hoped for, is from the police who have been alerted by a public-spirited neighbour or passer-by. The weak link is that it could be some minutes before the alarm is noticed and the police informed, and further minutes before they arrive. Where there are high-value goods, daring intruders may take a chance on this and stay just a few minutes to grab as much as they can before disappearing.

Therefore, the sooner the police can be informed the better. At one time it was possible to have a private telephone wire to the local police station, and a signalling system would send an alarm on being activated; faulty or broken telephone wires were also indicated making the system virtually foolproof. Unfortunately, the high incidence of false alarms has caused this facility to be withdrawn. It is still possible to summon help by automatic telephoning either by telediallers or digital communicators, though it is less effective than was the direct wire system.

Telediallers can be analogue or digital. With the analogue ones, a telephone number is recorded on an endless tape loop followed by a verbal message. When activated by the alarm system, the tape starts and the number is dialled, then the message is delivered. The system is not unlike a telephone answering machine except that it does the dialling instead of waiting for a call. Several different numbers and messages can be sent in succession, so a 999 call can be followed by calls to responsible staff members.

One advantage over the direct wire system, is that a special line is not required, the normal telephone line can be switched over to the dialler after working hours. The snags with these devices are first, unlike the answering machine you cannot record your own message, it must be done by the company that supplies the equipment. This is because the telephone number to be dialled has to be specially recorded using tones that the telephone exchange will respond to, before each message. If any changes are required, which may be occasioned by changes of staff or telephone numbers, the whole thing must be re-recorded.

The other drawback is that the machine dials each number once only. If it does not get through first time it does not try again. However, if several numbers are dialled, the chances are that at least one will be contacted. It should be noted that in some areas, police will not even accept calls from telediallers due to the high rate of false alarms. It is indeed unfortunate that what could be a valuable security aid is thus rendered unavailable by carelessness in installation or operation of alarm systems, and underscores the high priority which should be given to preventing false alarms. If such a police refusal operates in your area, at least staff can be alerted by the telediallers and can take appropriate action.

Some telediallers are dual purpose. A separate tape can be actuated by fire detection devices and a 999 call sent to the fire services, as well as additional calls to other numbers.

The second type is the digital dialler. This also makes use of a pre-recorded tape to send a suitable message, but the 999 number is contained in a digital memory instead of on tape. In this it is like those telephones that have a memory for storing frequently used numbers. The 999 call is the only number it will dial, but the message of some half a minute duration, will be continuously repeated for up to four minutes. This overcomes the possibility with the analogue dialler that part of the message may be lost due to noise on the line, momentary distraction of the one taking the message, or other causes.

An alternative to both analogue and digital diallers is the digital communicator. These do not use tape at all and so are more reliable; they are also cheaper. When activated, they dial a computer at a central receiving station which sends back a 'handshake' signal. If this signal is not received by the communicator it assumes it has not got through and so keeps dialling until it does receive it. This is an improvement on the diallers which give up after the first attempt because they have no means of telling whether contact has been made.

When the return signal is received, the communicator sends a digital code which identifies the sender, then transmits a coded message. The computer finally sends back an acknowledgement signal. If this is not received, the communicator redials and repeats the process. Most communicators can send only one type of message, but others can send different codes to correspond to different situations: an intrusion, personal attack (actuated by a panic button), or fire. Some can send up to eight different codes.

Further facilities offered by some models, are the dialling of two telephone lines to the same control station to ensure a connection, and the dialling of auxiliary numbers in addition to that of the receiving station.

On receiving an alarm signal, the station contacts the police and any other party designated by the subscriber. Even with these, there is some reluctance on the part of the police to respond as a result of so many false alarms. They put the onus on the station to filter them out, but the station has no means of identifying a false alarm from the genuine ones. So their response is to refuse to accept subscribers whose alarm system has not been installed by an approved installer. This cuts out the DIY installation.

Receiving stations are run and maintained by private security firms, and also British Telecom, most areas have at least one. These are generally advertised in the Yellow Pages as a 24 hour central station service. Payment is usually a fixed annual subscription plus a charge each time the service is used.

Diallers and communicators often have the facility of controlling the sounders that are operated by the alarm system. In particular, this means delaying the sounding of the bell for a set period to enable the police to arrive and catch the culprits. If a fault is detected on the telephone line they switch to instant sounding. As pointed out earlier, delayed sounding is a questionable practice as thieves could make a valuable haul or cause much damage and be gone before the police turned up.

There are some circumstances where this is not so. In premises such as a bank, penetration of the perimeter defence triggers the alarm system including the dialler, but further time is needed by the intruders to gain access to strong rooms or safes. There is thus little danger of allowing them to remain undisturbed for a short while, so that they can be caught. There could be other situations where similar conditions apply. It really depends on how portable the merchandise is, and whether there is immediate or easy access to it on entering the perimeter.

Modern digital communicators are quite small compared to the tape dialler. Many are just a printed circuit board that can be installed inside a control panel, and some control panels already have them fitted.

The details to be transmitted are contained in an electronic memory device called an Erasable and Programmable Read Only Memory (EPROM). What this mouthful really means is that it is a memory chip which data is read from, but to which data cannot normally be added by the user. The data is placed in it (it is programmed) by a special device which can also erase and reprogram it if required. This is done by the makers or the suppliers.

Radio pager

Another option well worth considering, is that of the radio pager. This is a small transmitter that is triggered from the alarm system, or it can be triggered from an independent PIR detector in the grounds of the protected premises, thereby giving warning of prowlers before any break-in occurs, or by any sensor guarding a sensitive area.

One example of this type of device, operates from a 12 V supply taking 50 mA (0.05 A) when on standby but 1.6 A when active. This is more than the usual control panel output, so an extra relay is required. It is supplied complete with aerial, and will transmit up to 4 miles. Remember this is 4 miles 'as the crow flies' and is further than a 4 mile journey by road. Hilly terrain or nearby high-rise buildings may reduce the range.

The receiver which has a pocket clip is a small breastpocket sized unit that flashes a light and sounds a bleeper when activated. It is powered by two small alkaline batteries, but rechargeable NiCad units could be fitted. Any number of receivers can be used, but they must all be tuned to the same transmitter frequency. A number of different frequencies are employed to reduce the possibility of one transmitter activating other receivers in the same area.

The device thus informs appropriate personnel of intrusion without the expense and problems associated with telephone links; also they can be reached when absent from home providing they are still within range. The disadvantage is that at ¾ in thick they are just a shade bulky to carry during an evening out in a dress-suit breastpocket. Perhaps on such occasions someone else can be persuaded to take the responsibility.

7 Closed-circuit television

Closed-circuit television (CCTV) may be thought to be prohibitively expensive, but the types used for surveillance work, especially the cameras, are much simplified compared to the video cameras used for video recording and entertainment, yet give surprisingly good results. So the cost, although not cheap, is not prohibitive even for a small business, and a basic system can compare favourably with many other types of property protection devices. Some firms specializing in CCTV equipment can supply second-hand and reconditioned items that reduce costs even further. Like any other system CCTV has its own features that make it more suitable for some applications than others.

Television is of course a highly technical medium, but the user of a CCTV system does not need to know all the technicalities involved. Familiarity with some of the basic principles though, can help to decide whether it can be beneficially used in any given situation, and just how it can be applied.

The camera

There have been big advances with the CCTV camera. Until recently, these used a vidicon as a pickup device which looks like a small cathode ray tube. As some of these cameras are not all that old and some may still be in use we will briefly describe them.

The picture is focused on the target of the vidicon which is coated with antimony trisulphate, and converts it into an electrical signal. There are two sizes commonly used, the 1 inch (30 mm), and the $\frac{2}{3}$ inch (17 mm), the latter being the more usual.

Definition is described as being of so many lines. This is sometimes confused with the horizontal lines that make up a TV picture but the two are quite different. The number of horizontal lines is a fixed standard which in the UK, Europe and most of the world is 625, the same as TV and video. In the Americas, Japan and a few other places the line standard is 525.

The definition of a camera or video system is defined in vertical lines and it describes the number of lines that the system will resolve without them running into each other and blurring. Broadcast TV has a definition at present of 300 lines plus, and video camcorders resolve around 240 lines.

CCTV cameras have a resolution of 500 lines and so should give detail better than even broadcast TV. This is because the output of the camera goes directly to the monitor without any processing, whereas that of the broadcast is processed and modulated onto a radio-wave causing loss of resolution. Digital TV though gives a much improved resolution. If the CCTV signal is recorded, as many are, then losses will take place, which, depending on the mode of recording, can greatly degrade the picture.

No new cameras now use the vidicon, but a pickup device consisting of silicon diodes called a charge-coupled device, usually abbreviated to CCD. These have been used in camcorders for some time, where lightness and portability are essential. CCTV cameras using vidicons were the industry standard and were much cheaper; their need for more complex power and scanning circuits made them larger and heavier, but this was no problem for security cameras. However, the cost of CCDs has tumbled and so they have been adopted as the standard CCTV camera now. A description of how they work is given in the Technical Section.

Sensitivity is measured in lux. Sometimes foot-candles are encountered but the conversion is simple, 1 foot-candle = 10 lux. A typical vidicon camera sensitivity is 20 lux. Some idea of how bright this is can be gained from these examples: bright sunlight gives around 100 000 lux; an overcast day some 5000 lux. Dusk, about 10 lux; bright moonlight 0.3 lux; half-moon 0.1 lux and starlight 0.001 lux. Ambient light in a city resulting from street and other lighting reflected from clouds is around 0.1 lux.

For all daylight and most indoor situations the sensitivity of the camera is ample, but outdoors at night could create a problem. This is an area where the CCD scores over the vidicon. Special low-light vidicons with trade names such as Newvicon, Saticon, and Chalnicon gave sensitivities down to 2 lux, but the CCD has a normal sensitivity of 0.1–0.5 lux. Some will go as low as 0.01 lux. So only where there is virtually no light will there be a problem, if the high sensitivity devices are used. Furthermore, the latest cameras can be switched between colour and black-and-white. In the later mode, the sensitivity is greater. Most CCTV cameras are now colour, whereas the vidicon cameras were all black-and-white.

In bright sunlight, sensitivity will most likely be too great, resulting in saturated colours, dense blacks and bright whites, but little detail. Highlights are absent because all is a uniform white. Thus faces appear as a featureless blob, which is the last thing needed if there is to good recognition and identification.

A manual sensitivity control is often provided which can be set to give the best results in situations where the lighting is constant, such as indoors with artificial lighting. With remote cameras working outdoors in varying lighting conditions, constant adjustment is obviously impractical. Instead, an automatic sensitivity control circuit is incorporated, and the sensitivity varies according to the light. This is not without its snags as we shall see, but it does give a generally well-defined picture under a wide range of lighting conditions. Control range can be from 10,000:1 to 100,000:1.

Another means of adjusting to the available illumination is by means of the lens aperture. Cameras are often supplied without lenses, and there is an extensive range to choose from, for the particular application intended. Most lenses for surveillance work have a standard screw-in fitting which is known as the *C-mount*. Different lenses are required for 25 mm and 17 mm tubes. These can be interchanged in some cases but will give a different focal length from that specified, and there may be corner shadowing.

Lenses for surveillance work are usually of fixed focal length, although various values are obtainable. Small focal lengths give wide-angle pictures, while large ones give close-up effects with a restricted width of view.

A focal length of 4 mm gives a field of view of approximately 95°, which is wider than a right-angle, and if mounted in a corner can encompass the whole area of a large square or rectangular enclosure, but the detail is small unless a large viewing screen is used. An 8 mm lens gives about 56° and is a good general size. The 16 mm lens offers a field of view of some 28° and is also commonly used where a wide angle is not required but greater detail of more distant objects is desirable. Larger focal lengths are also available to give an even closer view.

Zoom lenses can be fitted where a general wide-angle view is required with the ability to close up for greater detail in any portion (Figure 44). Each lens has a specified focal length range, a common one being 12.5–75 mm. This is a ratio of 6:1, which is thus the range of the zoom. An object seen at the minimum focal length would thus be magnified 6 diameters larger when observed at the maximum setting. Zoom lenses for CCTV range from 4:1 to 10:1.

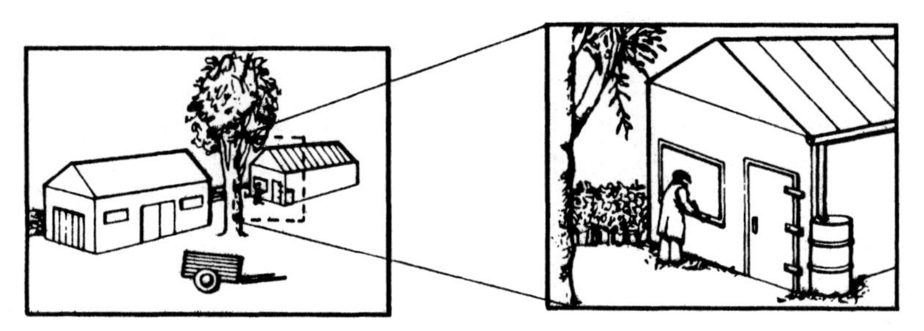

Figure 44 Wide-angle view for overall surveillance, with the actual effect of 6:1 zoom to examine any portion more closely. Zoom, as well as tilt-and-pan, can be remotely controlled.

For remote surveillance a motor-operated zoom is required, and also a motor-controlled pan-head to pan the camera sideways to the region needing examination. Alternatively, zoom lenses can be used at a fixed setting to exactly cover the required area and thus achieve optimum balance between coverage and magnification in any given situation.

Apertures for simple lenses are fixed, from *f*2.0 to *f*1.3 being the usual range. The smaller the *f* number, the larger the aperture and the more the light that will be admitted. Zoom lenses have variable apertures, the quoted number being the largest one, from which they stop down usually to *f*5.6. Most lenses have a variable focus and it is worth noting that when operating with a large aperture, the focus is more critical, with objects going out of focus more readily as they approach or recede from the camera. So always use the smallest aperture consistent with good contrast in the available lighting. Automatic apertures allow for light ratios of up to several million to one.

Worth noting too, is that lenses, especially zoom lenses are expensive, in some cases costing more than the camera. Costs become even greater if motor control is required, which it usually is with a zoom lens. If a simple fixed focal length lens will do the job there is little point in wasting money on a more expensive unit, and it may be cheaper and give better security to use two cameras to cover an area, rather than a single one with an elaborate zoom lens.

Monitors

The monitor is the term used to describe the display unit. It is similar to a TV receiver but has no tuning circuits and so cannot receive TV broadcasts. Unlike most TV sets the mains is isolated from the receiving circuits and so there is no possibility of connecting cables becoming live.

It is usually housed in a robust metal case which is intended to stand up to rough usage, and the cathode ray tube which displays the picture and its associated circuits are designed to give much better resolution than the domestic TV receiver. Over 700 vertical lines are typical.

Monitors are available in screen sizes from 9 in to 24 in measured diagonally, the most common being 9 in and 12 in. As with TV receivers, there are controls to adjust brilliance and contrast, while others, usually pre-set, are provided for the horizontal and vertical hold, height and width.

Distribution

The output from the camera is a video signal of approximately 1 V peak-to-peak. It contains very high frequencies of up to 10 MHz (10 million cycles-per-second) and so requires screened coaxial cable the same as used for TV aerial downleads. If there is a powerful transmitter nearby such as an ambulance or taxi base station, cable with a double screen may be required to avoid herringbone patterns appearing over the picture.

A number of monitors can be run from one camera. Each monitor has two signal sockets, usually of the bnc type. The cable from the camera is connected to the one, and another cable going to the next monitor is connected to the other. From the second monitor, a further cable is taken to the third and so on. Thus the monitors are looped from each one to the next (Figure 45).

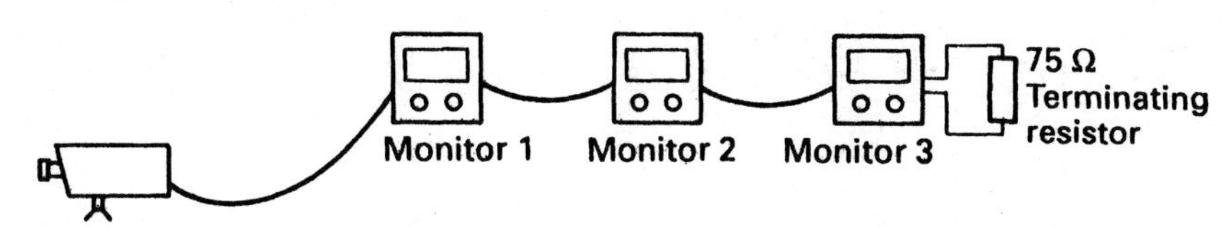

Figure 45 Video signal is conveyed to monitors via coaxial cable. Where more than one monitor is required, one must be looped from another, the final one being terminated with a 75 Ω resistor.

One technical consideration is important and must be taken into account although it is not necessary to fully understand it. Fortunately, its application is quite straightforward. The output of the camera is at an impedance of 75 Ω. This is similar to the output impedance of a hi-fi amplifier which may be 4 or 8 Ω. In the case of the audio amplifier, matching is not critical and an 8 Ω speaker can be connected to a 4 Ω output (though not the other way round).

With the CCTV distribution system everything must match, and that includes the cable. Coaxial cables have different *characteristic impedances.* This is not the same as the cable *resistance* which increases with length, the impedance of a cable is the same for any length (see the Technical Section for a detailed description). Cables are commonly available in 50 Ω and 75 Ω versions, so the 75 Ω type is the one to be used.

If two similar impedances are connected in parallel, the resulting impedance is halved, if three are paralleled the value is reduced to a third and so on. To correctly match the circuit, the impedance of the monitor should be 75 Ω, yet this would mean that only one monitor could be connected, as adding others would reduce the impedance below the required value.

To get round this problem the actual input impedance of the monitor is made high enough to have negligible effect on the circuit, but each monitor has a 75 Ω resistor which can be connected via a switch across the line. All are switched out except the last one. Thus the correct 75 Ω load appears across the end of line irrespective of the number of monitors used.

If the line is left unterminated by the switched-in resistor, energy is reflected back along it instead of being totally absorbed. A video signal will thus arrive at earlier monitors fractionally later than the original and produce a ghost image to the right of the true one. Sometimes the signal is reflected back and forth along the line several times before being absorbed by cable losses, resulting in several ghosts of diminishing intensity.

Cables introduce losses which reduce the video signal, but up to 650 ft (200 m) can usually be used before significant degradation of the picture occurs. If longer runs than this are necessary, low-loss coaxial cable should be used.

An alternative cost-cutting method of distribution especially if several monitors are required, is to use a modulator to convert the video signal from the camera to one similar to the broadcast TV signal. Then instead of CCTV monitors, black-and-white TV receivers can be used which can usually be picked up cheaply second-hand. The modulator is quite inexpensive.

The snags with this are that second-hand TVs are not as reliable as professional monitors; they are liable to go off tune and need re-tuning; and the cable losses at the high uhf television frequencies are much greater, so the system will work satisfactorily only over short distances.

Large area surveillance

If the area to be observed is greater than can be covered by a single fixed camera, or areas in different locations need surveillance, there are two possibilities. The first is to use two or more cameras mounted in suitable positions. These can either be connected to individual monitors, so that all areas can be viewed simultaneously, or they can be switched to a single monitor via an automatic sequential switcher. The amount of time spent on each camera can be adjusted with these from a few seconds to a minute, and the sequence can be halted at any scene if required.

Some switchers permit the connection of two monitors, one displaying the automatic sequence, while the other permanently displays any one picture selected manually. Thus activity in any one picture can be scrutinized while the other scenes continue to appear in sequence on the other monitor. This frustrates the ploy of causing a diversion in the range of one camera while the real attack takes place elsewhere (Figure 46).

A single camera can be used to cover one large area if it is fitted with a motorized pan-and-tilt head. Panning refers to horizontal camera movement, while tilting describes vertical movement. If the camera needs only to scan at one level which very often would be the case, a pan-only head could be used, which is less costly than the pan-and-tilt variety.

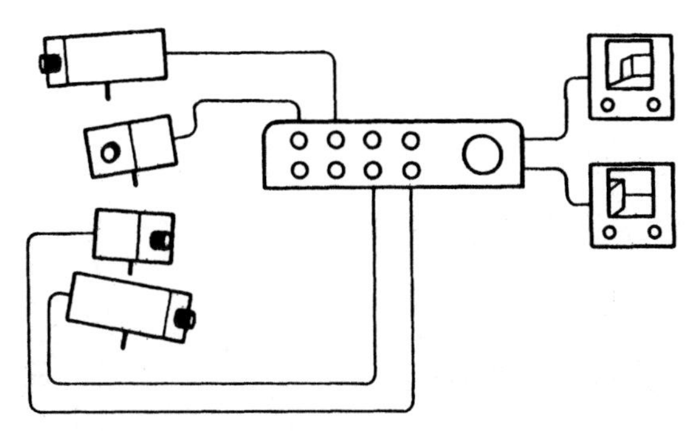

Figure 46 CCTV system having four cameras, a sequencer and two monitors on which one displays the four scenes in sequence and the other any one selected.

Automatic surveillance

Obviously there must be someone on hand to view the monitors continually. This is no problem during working hours in stores and shops, where the monitors can be sited to be in the view of security or management staff. After hours though, normal CCTV equipment is of little use unless round-the-clock vigilance is maintained by a security guard.

It can still be effective, however, if used with a *motion detector*. With this device, a number of points on the picture are selected by movable small circles or rectangles which appear on the screen. These can be electronically positioned to cover vulnerable areas. If any movement occurs in any one of these, the video waveform will change at that point, and such a change will trigger an alarm, either the main security alarm, or a special independent one.

Movement elsewhere in the picture has no effect. Thus the device is inherently resistant to false alarms and can be used to protect areas quite close to places where there may be legitimate movement. A further feature with some models, is that the minimum size of moving object required to initiate an alarm can be set. Small animals or birds can thereby be ignored, and another possible cause of false alarms eliminated. Alternatively, where there is no risk of small intrusions such as indoors, a larger area can be sensitized to give a higher degree of security.

Phone line television

This is widely used in America, but so far as is known, there is no UK manufacturer of such a system. A description is included here because new equipment is appearing all the time, and a UK version may well appear on the market in due course.

The device is linked to a video camera, and it freezes a single frame just as a video recorder can, although the technical means of doing so is quite different. The picture is stored in an electronic memory which, when instructed to do so, transmits it down the telephone line as a frequency-modulated audio signal. At the other end it is decoded by a receiver and fed to a TV monitor.

With high resolution models the picture takes 35 seconds to transmit, but with lower resolution which is still quite adequate for security purposes, the transmission takes 8.5 seconds. The picture is thus updated every 35 or 8.5 seconds on the display monitor.

An ordinary audio tape or cassette recorder can be used to record the signal, and thus provide a permanent record of any intrusion for subsequent reproduction and possible identification of the intruder. With a stereo recorder, sound can be recorded on the other channel at the same time and so combine an audible with a visible record of the event.

The telephone link can be dedicated to give permanent connection and display, or the system can use the normal line by using a dialler triggered by any suitable sensor set off by the intruder. The recipient of the call would recognize its origin by a code and immediately switch it to a decoder and TV monitor. This is obviously the cheaper method.

Features of the system are: the cause of the alarm can be seen, thereby saving calling out the police if the alarm is false; possible identification of the intruders and the ability to pass a description to the police to aid apprehension; and the recording facility for future reference or identification. The main drawbacks are that someone must always be available to answer the phone at the receiving end, and the line could be engaged at the wrong time.

Recording

Video recordings can be made from any CCTV system and are useful for identification. They could also help track down the cause of an otherwise inexplicable false alarm. Any of the home video recorder systems can be used, as the line and frame characteristics for CCTV is the same as for broadcast and domestic units. Special security video recorders usually use the VHS system and have many speeds, the slowest allowing some 200 hours recording with superimposed time and date. They also permit playback at a faster speed and freeze-frame.

There are two approaches to the problem of the long periods when nothing is happening. One, adopted by some professional units, is a unit that runs continually at a slow tape speed with limited resolution. A refinement often included is to speed-up the tape and so increase resolution when activity is sensed by a motion detector or a conventional sensor is activated.

The other approach is to have the recorder in the quiescent mode so that it can be rapidly activated by a sensor or by the alarm system. It thus does not need a long tape to record endless hours of inactivity. This is also kinder to the video record heads which have a limited life and are expensive to replace.

If a domestic recorder is obtained for the purpose, it need only be of the most elementary kind. Many models bristle with a host of facilities such as multi-event timers, remote controls, two or more tape speeds, hi-fi sound, and elaborate editing, none of which is essential for security work.

A facility that is useful is a good interference-free freeze frame, as this permits clear still photographs to be taken from the screen to assist in identification. A recorder may be obtainable at low cost in the now-obsolete Beta format, as these are often sold off because of the non-availability of pre-recorded tapes. This of course is no drawback for security applications, and many of these are excellent machines.

Other facilities and features

Cameras intended for outdoor use can be provided with weatherproof housings, and there are housings for practically any environment. These include dustproof, heatproof, pressurized, underwater use, and some are even explosion-proof!

If detection is required in completely dark surroundings, this can be provided by the use of infra-red sensitive cameras and infra-red lamps. Another problem may be the providing of power for cameras that are situated at a distance from a mains supply. Some systems are designed to supply power for the camera from the monitor, via the video cable. There are also special amplifiers which allow a video signal to be sent along up to 3 miles of twin twisted flex, thereby enabling very remote locations to be kept under surveillance.

A camera can be temporarily 'blinded' by car headlights, sun reflections or other strong light shining directly into the lens. When these are sensed, the automatic sensitivity circuit reacts to reduce the sensitivity, and so the rest of the picture goes dark leaving just the glare of the light. A device that prevents this is known as an *eclipser,* which effectively removes the highlights from the picture. Thus the source of dazzle appears black with just a halo of light around it, while the rest of the picture is normal.

Further advantages of the CCD

The vidicon had other disadvantages beside the lower sensitivity. It could easily be damaged by strong light entering the lens. This necessitated careful positioning so that the sun would not shine directly into it at any part of the day, which could easily be overlooked with outdoor cameras. A further factor that had to be considered was the position of the sun at different times of the year. Especially in the winter months, when the sun is low over a wide angle, direct sunlight could enter the lens when it would be quite clear during the summer.

For the same reason care had to be taken when transporting vidicon cameras, as damage could be suffered if inadvertently pointed toward the sun, even if the camera was not switched on or powered. A lens cap was therefore necessary. CCD cameras are not as vulnerable to such damage, though it is wise not to allow the sun at its brightest to shine in the lens for any length of time. A high elevation with a downward camera angle should avoid any such problems.

Another effect with the vidicon was the ghost image. If the camera was aimed at a well-illuminated scene for a long period, the bright portions of the scene would be burnt permanently into the screen, and would appear as a background to all subsequent images. This does not happen with the CCD.

Flaring, that is the appearance of streamers behind moving objects, especially with dark scenes and rapid movement, was often observed with the vidicon. It could seriously affect indentification as faces were blurred by the effect. This does not happen with the CCD. It can be seen that the CCD which also allows smaller cameras to be used, is a big advance over the vidicon.

Applications of CCTV

Closed-circuit television is not suitable for all security situations, but is ideal for others. Domestic or small business premises are better served by ordinary alarm systems which do not need a continual watchful eye. But wherever there are are large rambling premises with outhouses, such as farms, market gardens, timber and goods yards and the like, CCTV can prove very useful to augment other security systems or to cover areas where conventional sensors could not give complete protection.

Security guards in large factory premises have often been attacked when on patrol. However, patrols can be reduced or even eliminated when CCTV is installed, with simultaneous surveillance being maintained over vulnerable areas from the safety of a central point in the building. Intrusions can then be immediately reported to the police, and if intervention by the guard is necessary, he is fully prepared and can see what he is up against.

CCTV is commonly used in stores, both large and small, to detect or deter shop-lifting. The deterrent factor is probably the most important, and to exploit this dummy cameras are often prominently displayed at various points. Some of these though look decidedly unconvincing, especially those having a multitude of 'lenses' sprouting at all angles from a dome-like casing. Many of today's rogues may be quite familiar with video cameras, due to the boom in home video systems, and will not be taken in at all by crude imitations. If dummies are to be used, make sure they look right and have video and mains cables coming from them.

Assisted by staff, goods often disappear through the loading bays at the rear of the premises. A real camera here with a monitor permanently on in the main office could prove a strong deterrent. This subject will be covered in more detail in later chapters.

8 General security

The installation of a well-designed intruder alarm system with or without closed-circuit television (CCTV) will go a long way to increase security by deterring would-be intruders, and scaring them off and warning neighbours should a break-in occur. However, reliance should never be placed solely on an alarm system however good it may be. A building with poor physical security may tempt a burglar in spite of the alarm, in the hope of snatching a few valuables before making a quick getaway, or possibly even silencing the alarm.

It is surprising the number of people who install quite efficient and expensive alarm systems and yet neglect basic security measures. The alarm system should be regarded as the second line of defence. The first defence is a solid and impenetrable (as far as it can reasonably be made) perimeter.

The front door

The average lock used on most front doors whether domestic or business, is one of the easiest things to open by even a semi-skilled thief. It can be sprung open by means of a thin piece of plastic such as a credit card pushed against the latch between the door and frame – an alternative criminal use to the more usual credit card fraud! If the door has a glass or thin wooden panel near the lock, it can be broken and a hand inserted to turn the lock from the inside.

Many locks have the staple (the part on the frame that the bolt engages with) fixed by short screws into the wood. A good shoulder charge or levering with a jemmy can easily loosen these and force them out.

Any such conventional locks or *springlatches* as they are more accurately termed, should be immediately replaced with a *deadlock*. This is a lock with a bolt that cannot be retracted without a key, that is, one that you cannot slam shut but must be locked with a key when leaving. It cannot be sprung open with a card nor can it be opened from the inside without the key.

The latter feature not only prevents the door being opened through a broken panel but has another important advantage. However burglars enter a building, they first seek to open an easy exist route through which to remove their loot and also effect a quick escape if need be. Inability to do so makes them feel trapped and limits the amount and size of goods they can remove. A deadlock prevents such an exit route through the protected door.

Most springlatches have a two-lever mechanism and if other means of opening it fail, can be picked without too much trouble. Security deadlocks have a minimum of five and some up to ten levers which is virtually impossible to pick. Another desirable feature is the provision of steel rollers set inside the bolt. Any attempt to saw through the bolt with a hack-saw is futile.

It is generally held that a mortice lock, i.e. one that is fitted in the door rather than screwed to its surface, offers greater security (Figure 47). This is because it cannot be removed from the inside by an intruder who has gained access at some other point and wishes to establish an escape route, nor can it be burst off the door by force.

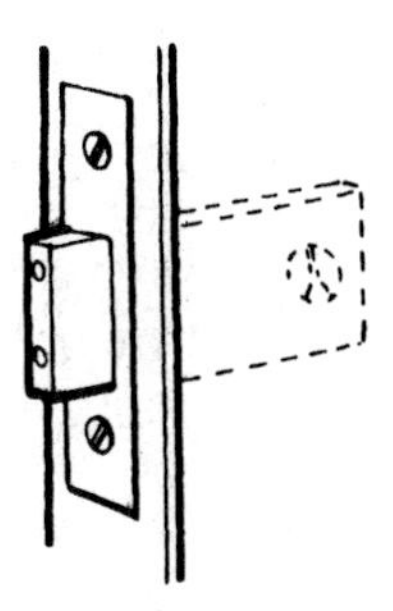

Figure 47 Mortice deadlock. This should be fitted only when door is thick enough to avoid weakening it when wood is excavated.

However, this needs qualification. Wood has to be removed from the door to make a cavity in order to fit the lock. Thus the door is weakened at that point. If the door is thin, as many modern ones are, there may only be a thin shell of wood enclosing the lock. The application of force could splinter it away rendering the lock useless. In addition, the staple is let in to the door frame, and the same weakness could exist here if the frame is insubstantial. So rather than improving security a mortice lock could considerably reduce it.

Some modern mortice locks are made especially thin to overcome this problem and less wood has to be excavated. Even so, for a thin door, the value of a mortice lock is dubious. It could be added that a thin door in itself is a security hazard and should really be replaced by something more substantial.

If there is some doubt about the matter, it would be better to fit a security surface lock, such as a deadlatch, which offers better protection (Figure 48). Bolts from the keyplate at the front of the door pass right through the door

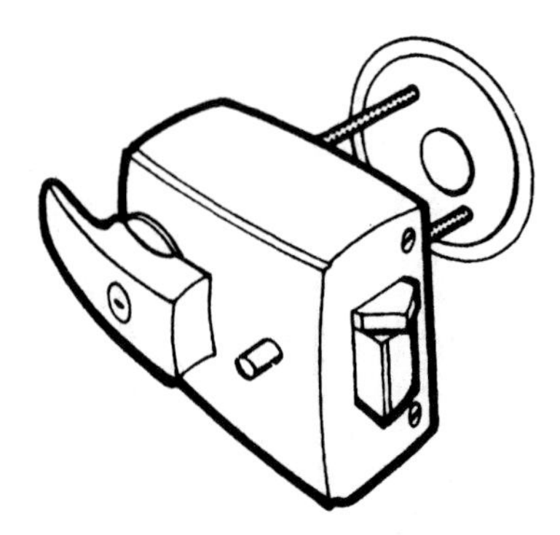

Figure 48 Surface deadlatch, high security type. This is fixed by screws at side and bolts through door from front keyplate, and can be double-locked to prevent opening from inside without the key. This type can be fitted in place of many ordinary springlatches without modifying or cutting the door.

to the lock on the back, and woodscrews enter the door sideways through the side flange. The lock is thus impossible to remove from the outside or inside while the door is closed. As it exerts a clamping effect on the door it strengthens rather than weakens it. An attempt at forcing an entry is thus unlikely to succeed. The staple is also secured by sideways screws which are concealed when the door is closed and will resist a considerable amount of force.

The type of deadlatch illustrated does not need to be locked shut but can be slammed just like a springlatch, however, the shape of the latch and its action makes it very difficult to spring open with a card. Another feature is the double-locking facility. The lock has a keyhole on the inside as well as the out. When the inside is locked the handlever is immobilized so that it cannot be opened from the inside.

One trick that has been used by intruders, is to spring the door frame apart with a car jack, at the point where the lock is fitted. Often it can be bowed sufficiently to disengage the lock bolt from the staple or rebate plate. To minimize this possibility the frame should have solid support at the sides. Weak materials such as plaster should be excavated and replaced with concrete especially near the locks, on both sides of the door, and any gaps should be filled with the same material. The rebate itself should be not less than ¾ in (19 mm) thick.

Further security can be achieved by having two locks spaced well apart. This adds to the inconvenience of locking and unlocking and means an extra key, and so is one of those decisions that has to be made as to the degree of security considered necessary. We shall be looking more closely at this in the next chapter.

Simple draw-bolts placed at the top and bottom of the door effectively defeat door-frame springing, but obviously cannot be used on the exit door. They can, and should, be used on all exterior domestic doors at night.

Most doors swing inwards, but any exterior door opening outwards is vulnerable to attack to the hinges, as the hinge pin is exposed on the opening side. It is not too difficult to remove the pin whereupon the door can be simply lifted away. Fire doors and those fitted to some outbuildings are usually outward opening, so these constitute a major security hazard.

The solution is quite simple, and takes the form of what are known as hinge or dog bolts (Figure 49). These consist of a recess plate and engaging lug which are fitted to the frame and door respectively on the hinge side. A pair should be fitted both to the top and bottom of the door. When the door is closed, the lug engages with the recess, so that the door cannot be lifted out if the hinge pin is removed. Even with inward opening doors, dog bolts can be fitted to reinforce the hinges against a forced entry, especially if there is some doubt as to the strength of the hinges. The beauty of them is that once fitted they need no further attention.

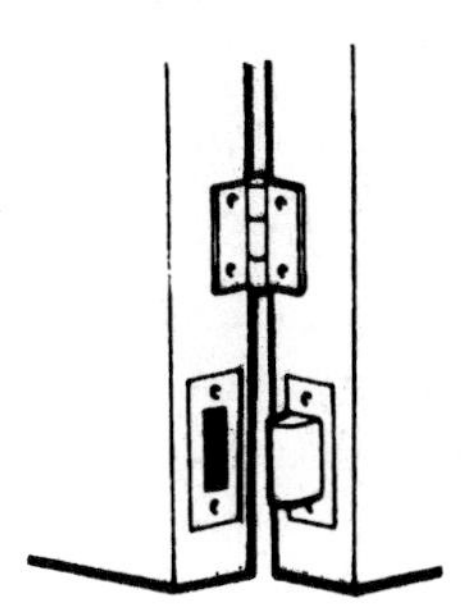

Figure 49 Dog (hinge) bolt. Interlocks door and frame on hinge side when door is closed. Prevents removal of door if hinge pin is removed with outward opening doors. Two should be fitted, top and bottom.

Where very high security is required, even a stout wooden door may be insufficient. Battering rams have been used to break down doors. A wooden door can be reinforced by covering the outside surface with mild steel of 16 gauge or thicker. The edges should be turned over and secured to all four door edges by rows of countersunk wood screws.

In addition, coach bolts should be fitted at intervals of not more than 9 in (225 mm) apart, the heads on the outside, and passing through the stiles and rails (the main vertical and horizontal door members). On the inside large washers should be fitted underneath the nuts over which the bolt ends should be burred.

A good quality mortice deadlock of at least five levers should be fitted, together with mortice lockable bolts if the door is not to be used as the final exit. If it is, then at least one other similar lock should be fitted. The increased weight will require an extra pair of hinges, and dog bolts should also be fitted on the hinge side.

This will give an extremely attack-resistant door, but do not overlook the frame which could now be the weak link. In addition to seeing that it is well supported by concrete at the sides, a strip of angle iron screwed to the frame at the opening side will prevent the insertion of a jemmy to force the door and frame apart.

Rear entrances

Many owners make the mistake of fortifying the front of their premises to a high state of impenetrability but have only the most rudimentary protection at the rear. This is in spite of the fact that a break-in at the back of the premises is far more likely. Burglars are well aware of this quirk which suits them well because they much prefer to enter at the rear. It is usually less public than the front, so they stand far less chance of being observed.

Rear doors should be of substantial construction and fitted with deadlocks plus lockable bolts top and bottom. These should be secured at all times, even when the premises are occupied, unless access is needed for loading or other essential purposes.

The same principle applies here as that for the front door, which is that doors should be secured so that they cannot be opened from the inside without a key. This blocks a possible escape route and easy passage for the removal of goods. Often, simple draw bolts are fitted and keys left in the locks for convenience, in the belief that their only function is to keep burglars out. Things are thus made much easier for the intruder who has gained entry elsewhere into the premises.

The lockable bolts we have mentioned are commonly known as *rack bolts*, and they can be either mortice or surface fitting (Figure 50). A common key fits any bolt of the same type and so enables a number of rack bolts to be used without the inconvenience of needing a key for each.

This means that they are not by themselves a high security device, and should always be used in conjunction with a deadlock. However, the intruder would only encounter them when he is inside and it is unlikely that he would have a suitable key with him. The greater the number of devices securing a door the stronger it is, and the harder and more trouble it is to break open.

Mortice rack bolts require only a small hole in the wood to receive them, so there is little weakening of the door. A version designed for windows is shorter than the door type so that it can be accommodated in a narrow

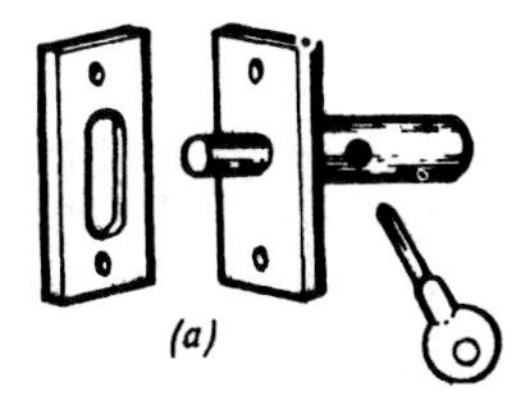

Figure 50 (a) Mortice rack bolt, with key. This needs only a hole drilled in door or window. The window type is shorter. Another smaller hole is required to insert the key.

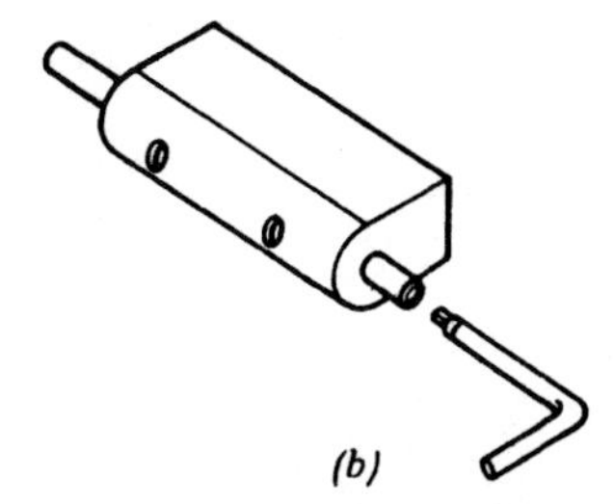

Figure 50 (b) Surface rack bolt. A key is inserted in the end to operate. All fixing screws are concealed. This type is used where a door or window has insufficient wood for mortice type.

window frame. A second smaller hole intersecting the first at right angles is required for the key.

Surface rack bolts are fitted when it is not possible to use a mortice bolt. All the screws are concealed once the device is fitted, and it has the advantage that it can be bolted manually without a key, although of course the key is necessary to withdraw it.

Windows

The majority of all entries are made through windows as these are the most vulnerable points and are usually the most neglected. Often windows are left open thereby offering a clear welcome sign to intruders. Ground floor windows are his first choice, but first floor windows and fanlights are by no means out of bounds to him. It should be noted that while double glazing is undoubtedly a deterrent, it is not burglar proof as is often claimed.

Few burglars will attempt to get through a broken pane of glass. There is too great a risk of being badly cut, unless all the broken glass is removed from the window frame, and this would take too long as well as itself being risky. Several methods are commonly used to open a window and gain entrance. One of these is by manipulation of the catch from outside. With a sash window, this can often be done by sliding the ubiquitous credit card between the frames and slipping the catch back. Another method where a small fly window has been left open, is to drop a looped cord inside and engage the main window handle which then can be easily lifted by pulling on the cord.

A common method is to break the window and insert a gloved hand to open the catch. In a refinement of this method, a few strips of sellotape are first stuck across the window to stop the pieces falling and making a noise.

Weathered and broken putty is a major security hazard as it can be quickly chipped away and the window glass lifted out. Then the catch can be released and the empty frame opened; this method is frequently used. Another is to prise the window away from the frame with a jemmy.

As all these modes of entry, except pane removal, depend on the window being opened, security can be achieved by preventing it from opening, providing the frame and puttywork is sound. There are a number of ways of doing this.

The short mortice rack bolt already described is one. This is ideal for the sash window, in which two bolts should be used, one on each side of the frame. These can be fitted so that the window can be locked when opened a few inches to allow ventilation.

Figure 51 Lockable window catch. This can sometimes be fitted in place of existing window catch. A locking catch should always be specified with new windows.

The conventional window fastening for a hinged window consists of a pivoted handle and catch, which can be easily released if the window is broken. These can be obtained with a built-in lock (Figure 51), and are now often supplied with replacement windows. These should always be specified for new windows, and it may be possible to fit one in place of

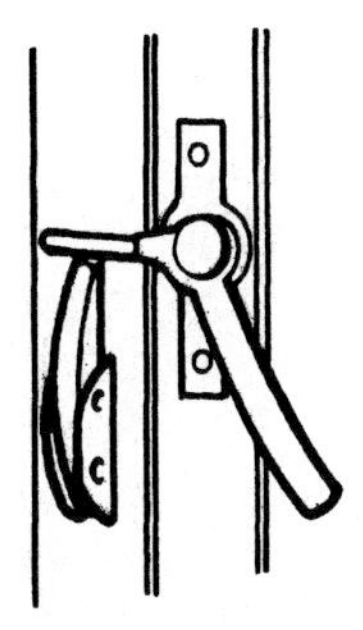

Figure 52 Window catch lock for fitting to existing catches. The arm swings up under the catch and is locked in place.

an existing non-locking catch. Those that are welded or riveted to a metal frame and so not easily removable can be secured with a locking arm that holds the catch in the closed position (Figure 52). A similar device can be obtained that secures the window stays that are commonly used for fanlights.

In many business premises there are windows that are never opened, such as those in corridors, store-rooms and other locations where no-one is actually working. Other areas have sufficient ventilation from frequently-used doors or forced-air heating and ventilation systems. The best security for these windows is to permanently seal them by simply screwing long wood-screws at an angle through the window to its frame. Sash windows can be screwed to each other.

In areas or with businesses of high risk, the possibility must not be overlooked that a pane may be removed even though the putty is sound, and an entry made without opening the frame. This would defeat window locks and also any alarm contacts fitted to the window (although not a foil strip which would be broken in the process). An expert could use a glass cutter to cut the pane close to the frame, then tape strips of foam plastic over the remaining glass edges to avoid injury when climbing through.

The only solution to this apart from alarm sensors in the room, is a window grille. For highest security, bars should be fitted across vulnerable windows. These should be of mild steel, not less than ¾ in (19 mm) in diameter, and set no more than 5 in (125 mm) apart. If they are over 20 in (500 mm) in length, tie bars should be fitted between them. The ends should be buried in the stonework to a depth of at least 3 in (75 mm).

This may seem a rather drastic treatment and make the place look like a prison. Expanded metal grilles may be visually more acceptable, and when fitted outside, also gives protection against broken windows caused by vandalism. They should be fitted to a steel frame and grouted into the stonework.

An alternative, which although giving slightly less security is still very secure, is scrolled ironwork fitted inside the window. Most towns have firms that will make this sort of thing to measure. The inside measurements should be taken very accurately measuring all four sides, then subtracting $\frac{1}{16}$ in (1.5 mm) to allow for warped or non-parallel surfaces and painting.

The makers will construct a frame to these measurements, drilling holes for the fixing screws, then fill the frame with scrollwork to your pattern. An example is shown in Figure 53. The scrolls are welded to the frame and to all points where they contact, making a very rigid and impenetrable barrier.

Figure 53 Example of scrollwork that can both protect and enhance a vulnerable window.

The grille should be painted with a rust-inhibiting paint before fitting, and long screws used to secure it to the window frame. If white or a light colour is used, the grille can look very attractive and actually enhance the appearance of the window rather than detract from it. It is visible from the outside and so deters attempts to break in.

It should be mentioned here that some DIY aluminium windows that were popular a few years ago are a considerable security hazard as they are screwed into the wooden surround by wood screws which are easily accessible from outside. A few minutes work with a screwdriver and the window can be lifted completely out. It is best to replace any of these with properly made units, but the windows can be made more secure by removing each screw and dipping the thread into liquid glue before replacing it. When set they will then be almost impossible to remove.

Louvred windows are also vulnerable as the louvres can be usually removed from the outside. These should be replaced, as little can be done to improve their security.

Fire safety

While it is necessary to secure windows from the possibility of intrusion, this should not be done to the extent that all escape routes are sealed if a fire should break out. First-floor windows are often the only means of escape for persons trapped in the upper storeys of a building, when fire engulfs the lower floor or the staircase. For all windows to be screwed fast or to have non-removable grilles could create a death trap.

Usually, the front first-floor windows, that is those facing the street are less likely to be entered by an intruder because of their height and risk of being observed. These are also the ones most easily reached by fire escape ladders. At only a slight extra security risk then, these should be easily openable from the inside. They can of course be wired with alarm detectors.

Security coatings

Access to upper storeys and flat roofs is often easily achieved by climbing drainpipes, posts and similar objects, while fencing and boundary walls are all vulnerable to scaling. Young vandals as well as burglars may be tempted to do so, and can cause just as much damage and loss.

An effective preventive is a special anticlimb paint such as manufactured by Camrex Special Coatings. It is applied as a one-coat layer, ⅒ in (2.5 mm) thick, over the surface to be protected. The treated surface is as difficult to climb as a greasy pole. It looks dry but stays soft and slippery for many years whatever the weather, and does not drip or sag in hot conditions. Any attempt to climb it covers the hands, shoes and clothes, in a clinging sticky mess. No doubt parents of youthful offenders would ensure that the attempt was not repeated! It contains an easily detected chemical that assists identification of suspects, yet it is non-toxic and does not permanently harm clothing. There is thus no risk of legal come-backs. But to avoid innocent inadvertent contact it is recommended that the coating is not applied to surfaces below 7 ft (2 m).

Graffiti is another problem that can cause heavy expense due to the frequent need for redecoration of exterior walls, toilets, corridors, etc. The same firm make a range of anti-vandal coatings which are said to be virtually impossible to mark with ball-point pens, lipsticks, crayons and felt-tip pens. Surfaces are impervious to aerosol paint sprays which can be easily removed with stain remover.

Various textures are available from heavy to smooth, some being coloured or random patterned containing multicoloured flakes, while others are clear to allow the original stone or brickwork finish to be seen.

Outbuildings

Outbuildings and sheds may contain little of value and so not warrant any special security. It is important though, to secure items such as ladders, crowbars, garden spades and the like, as these could be used to gain entry into the main building. Padlocks are usually used to protect such premises, but these like door locks differ in the degree of security offered. It is little use using a high security padlock with a fitting that can be simply unscrewed from the door, or from which the hinge pin can be punched out. Locking bars should be fitted that conceal all screws when closed and have countersunk hinge pins.

As well as being used to gain entry, ladders are a popular item for theft according to police reports. It seems they are used by builders and window-cleaners in the 'black economy'. They should never be stored outdoors, but be locked away inside and chained and padlocked to some structural member. Chains should have welded links, otherwise the link joins can be readily pulled apart, and should be substantial. Furthermore, the firm's name should be painted prominently along both sides of the ladder.

Safes

Safes of less than one ton should be fitted to the structure of the building. The makers can supply fixing devices for recent models, and older ones can have angle iron welded to the sides and top. It should be encased in not less than 6 in (150 mm) of reinforced concrete which is well keyed to the walls and floor.

Underfloor safes offer better security than wall models, and smaller units can be used. They should be buried in the floor in a hole with diverging sides, reinforcement rods fitted between the safe bottom and the sides of the hole, and then the space filled with best quality concrete.

In addition to the difficulty of attacking such a safe, its presence may not even be suspected by the intruder as the top can be easily concealed. If required, it can be fitted with a 'letterbox' facility, which allows money to be dropped in it by staff not having a key, but not permitting access to the contents. This is particularly suitable for petrol-station forecourt applications. Deposits of banknotes can then be made whenever the till accumulates more than a stipulated amount.

Personal attack

All these measures could have the effect of concentrating the attention of would-be felons to the weakest link, the staff member conveying the takings to the bank. A ruthless attack, perhaps with lethal weapons gives little hope of defence, nor would it be prudent to resist under such circumstances. Until such a time as our lawmakers see fit to restore the ultimate deterrent, thugs will not hesitate to kill in order to quell resistance and eliminate the risk of their future identification.

The best defence is uncertainty. Trips to the bank should never be made at the same time of day, or if weekly, on the same day. Random timings will confuse and discourage any who may be watching and planning an attack.

As with the security required for premises, the degree of security needed depends on the circumstances. The proprietor banking the takings of a small corner shop would not be at the same risk as staff conveying large amounts from a busy store, unless in a high-risk area.

Where risks are greater, in addition to staggering the times, different members of the staff could be used on different occasions, the cash could be carried in different containers, and different routes taken. Further confusion could be created by the routine sending of staff members carrying containers that could contain cash, on various errands as decoys, at various times.

High security cash-carrying cases are worth considering. These can take various forms. One type is made in the style of an executive brief case, another, like a lady's shoulder bag. Both have a wrist strap which breaks away if the bag is snatched. This activates a loud, piercing alarm, and also releases an emission of dense orange smoke which stains the money inside as well as the hands and clothes of the thief.

The same ploys can be used for the collection of wages. Where possible, wages can be made up from takings and thereby reduce the amount of cash in transit. Many firms pay wages by cheque which completely eliminates the risk. But employees have the right to be paid in cash if they insist, so some cash payments may still have to be made, thereby constituting a security hazard if sufficiently large. A firm may consider providing some form of incentive to encourage all to take their wages by cheque.

For large cash movements it is prudent to employ the services of a security firm, as it is then their responsibility to ensure safe delivery.

9 Planning a security system

As pointed out previously, security is relative. It is virtually impossible to make any premises absolutely invulnerable; bank vaults are probably the nearest thing to it, yet even they are broken into at times. The important thing is to achieve adequate security sufficient for the degree of risk.

For example, a jeweller's premises needs very high security because the high value and extreme portability of the goods can attract professional thieves who are both expert and prepared to go to a lot of trouble. Many of these have a knowledge of alarm systems and how they can be circumvented.

A shop selling TVs, videos and computers, also needs high security as these are popular goods of high value and easily disposed of. Usually though, these attract the local villains who use force rather than finesse, and they can be thwarted by measures that would not defeat the jewel thief. Security thus need not be quite so high.

Chemist's shops can be the target of drug addicts so good security is necessary, especially for the area where drugs are stored, but it can be of a lower order in other parts. Likewise the corner general store; here the attraction is cigarettes and loose change, and it is normally the target of teenaged opportunist thieves. A conspicuous alarm system and well-secured doors and windows will usually send them looking elsewhere.

Apart from the type of business, other factors can influence the degree of security required. The presence of a large council housing estate in the vicinity, a nearby public house, school, night-club or sports facilities can all increase the risk. Premises in dimly-lit side roads or those with no nearby houses or passing traffic at night are also more vulnerable. Those at the end of a terrace are at greater risk than premises within it, as are detached or semi-detached buildings. Lock-up shops are a more attractive target than those with occupied living accommodation. Also to be considered is the crime record for the neighbourhood. Some have a high incidence of crime while for others it is comparatively low.

Over-elaborate security measures, for the risk involved, can be a waste of money and make things irksome for all those who have to live and work with them. Yet even if all the factors seem to indicate low risk, it is wise not to be too complacent. Owners are becoming more security conscious, and premises are now better protected than they were, so those that have poor security are the ones that the criminals are looking out for.

So it is a matter of exercising good judgement. A good standard of general security with a reliable alarm system should be considered a minimum requirement, but adding the higher security features that have been noted in the foregoing chapters depends on the considered risk.

Become a burglar!

Well, in thought if not in deed. Try to look at your premises through a burglar's eyes, put yourself in his shoes, take a walk around from the outside to see where you would break in if you had to. If the cap fits uneasily, imagine instead that you have locked yourself out and you have got to break in somehow; but to save embarrassment you want to do it unobserved by passers-by or the neighbours. One difference though is that as a burglar you are young, possibly in your teens, are agile, able to climb about, and have a good head for heights.

This means that you should not overlook possible first floor entries, especially if there is a nearby wall, post, pipe, tree or any other means of getting up there. The burglar will probably be a lot slimmer than you are, so narrow windows or fanlights will be no problem to him. Some victims have expressed surprise that entries have been made into their property through, what was to them, unbelievably small apertures.

Remember that the intruder does not expect to leave by the same way; his aim is to open up an exit route through the front or back door, preferably the latter as he can then escape if he hears someone returning through the front. The importance of making all doors unopenable without a key can thus be appreciated. The intruder then has to face leaving the same way that he entered, so risking being seen, and which could be tricky if he had to do it quickly. Many a burglar has had a nasty fall doing just that.

The desire of the intruder to be unobserved means that points affording some cover are especially vulnerable. Trees, bushes, walls, advertising hoardings are all very welcome to the burglar. For the same reason, poorly lit areas away from street lights are his favourites.

Detached, semi-detached, or end-of-terrace premises are also preferred because there are neighbours on only one side, and so there is less possibility of being observed or heard. In addition, there is likely to be a freer escape route from a rear which is not hemmed-in by two neighbouring gardens.

After taking a criminal's eye view of your premises, note down the possible entry points in order of vulnerability. Ground floor doors and windows will head the list. If there are any outhouses at the rear through which access can be obtained to the main building, these are likely possibilities. Then comes the first-floor windows that can be reached by climbing. Flat roofs too are easily accessed and broken through. Do not omit any reasonable possibility, remember the old adage that the strength of any chain is in its weakest link. Leave just one point unprotected, and that will be the one through which the intruder will get in. Then all the work and expense on the rest will have been to no avail.

Before even thinking about an alarm system, get the physical security in good shape. That is the first line of defence and the first priority. We will take a look at some specific points for particular types of premises in the following sections.

Shop premises

Most shops are in a terrace with others in which case there are no possible entry points from the side. The rear is the most likely means of access, so all possibilities here need special scrutiny. In particular, check the rear windows carefully. Ensure all puttywork is sound. If never opened, screw them up, and if particularly vulnerable according to the criteria we have established, fit metal grilles. The rear door should be substantial, and if it has thin wooden panels as so many have, either replace the door, or screw and glue thick plywood pieces over the panels on the inside. Fit a deadlock and rack bolts at the top and bottom. If the door opens outwards, fit dog bolts to the hinge side at the top and bottom. In vulnerable or high security situations it does no harm to fit dog bolts anyway. For very high security, further rack bolts and metal cladding may be required.

Ensure that first-floor windows are sound and if at all accessible fit grilles to these too and screw them up. Unless used for living accommodation most rooms over shop premises are store rooms so there will be no need to open the windows.

Do not keep anything of value in outhouses, they can be used to store such things as advertising and packing material, and the odd accumulated junk which is not a lot of use but seems too good to throw out. If there is access to the main building through an outhouse, concentrate on securing the door into the building rather than the one from the yard into the outhouse. An entry may well be forced through the outhouse door, windows or even roof, they are not very easy to adequately protect. But if the door into the main building is well secured the intruder will get no further.

For maximum security, the perimeter wall of the back yard can be built up high and the entrance gate made to match, and kept locked. While the wall could be scaled, it presents another obstacle and another trap should a quick getaway become necessary. If it is never used for incoming goods or waste disposal, the gate could be removed and the space walled up unless fire regulations require a rear exit.

A passive infra-red (PIR) operated floodlight directed into the yard completes the rear external security measures. The low initial cost, low running costs and convenience of use makes the PIR floodlight an excellent security investment even where high security is not needed.

Coming now to the front of the shop or store, the window is vulnerable if expensive merchandise such as jewellery or electronic goods are displayed. Physical security can be obtained by means of a protective grille that offers minimum visual restriction; a night shutter – even a simple venetian blind would create problems for the quick smash-and-grab merchant; or the removal of expensive items each night.

Another possibility is double glazing with at least an inch or so separation. This is an expensive solution but though not totally secure, the obvious difficulty in penetration and removing objects through two layers of broken glass would undoubtedly deter attempts. Plus points are: no restriction of vision; reduction of the large heat loss that occurs through a normal shop window; reduction or elimination of condensation inside the window; and reduced traffic noise making for a quieter and more peaceful environment. The latter tends to give a 'quality' atmosphere which could have a beneficial effect on sales.

Large objects on display could be linked with chains to each other or to anchoring points in the window bay. These need not be heavyweight, as the thief would have no time to use a hacksaw. Speed is all important to the smash-and-grab thief so anything that hinders him, increases security.

Shop doors have often large areas of glass which are especially vulnerable. Scrollwork grilles offer good security as well as being decorative, so these are ideal for door glass protection. Two locks should be installed if anything more than moderate security is required, one of which at least should be a high-security deadlock with no less than five levers.

Do not overlook a trick that has been used on many occasions. That is the 'fishing' of articles from the counter, or display stands by means of rods through the letter box aperture. It is not so difficult as it sounds. A strong wire mail box behind the letterbox is the best answer to this one.

While first-storey windows over the shop front are not usually a high risk, they should not be overlooked. If they are opened at times, lockable catches should be fitted, but if not, it would be better for them to be screwed up. This should NOT be done if staff are working on, or the public has access to, the first floor, as they may be the only escape route in the event of fire.

Pharmacies

Chemist's shops are different from all others inasmuch as a certain minimum standard of security is required by law to protect the stock of drugs. The stipulations are laid down in the *Misuse of Drugs (Safe Custody) Regulations 1973*. These deal with the physical security of controlled drugs when stored at premises not covered in Section 11 of the *Misuse of Drugs Act 1971*. Schedule 2 of the regulations gives detailed specifications for a cabinet or safe required for drug storage in pharmacies, nursing homes and similar places.

However, regulation 4 allows a local Chief Constable to grant a certificate of exemption from the precise compliance with the specifications if he is satisfied on inspection, that the general security of the premises is sufficient to prevent the theft of drugs. The certificate is renewable annually. Not all police forces grant such certificates and at the last review, five out of the forty-three forces in England and Wales did not.

It follows that those who qualify for certification must have a high standard of general security, while that of those not certificated is likely to be lower. The latter of course must have drugs stored in a container complying with the regulation, but are not obliged to secure the rest of their premises.

Pharmacy burglary statistics show that 14.5 per cent of non-exempt premises suffered theft of drugs, while only 2 per cent of exempt premises did. This is clear testimony to the value of good general security.

Shop alarm systems

Physical security is essential as the primary defence; the alarm system is the back-up, always highly desirable, but for high risk situations, essential.

The precise details of a system will depend on the premises. However, there are general points common to most. Again, looking at the rear, any window that can be opened should have magnetic contacts, while non-opening glass should be protected with a foil window strip. Doors also should have contacts, all being connected to a detection loop with 24 hour anti-tamper wiring to warn of any interference with it by 'customers' during normal shop hours.

At the front, the window can be protected by sonic vibration detectors, inertia detectors or metal foil. In the case of a smash-and-grab, considerable noise is generated and any alarm sensor may be thought superfluous. However, if all adjacent shops are lock-ups and there is no-one about, the sound of breaking glass which is momentary, may not be heard. A continuous sounding alarm is a different matter and at least is likely to panic thieves into just grabbing a few items and getting away quickly.

To deter is by far the best objective here, as the cost of replacing the window is likely to be much greater than the items lost, to say nothing of the disruption caused. So, the physical security measures described earlier should be the principal mode of defence.

Of all the sensors, the metal foil is probably the best for conventional shop windows. It is simple, inexpensive and cannot fail. Furthermore, its presence is obvious and this itself serves as a deterrent. It should be connected to a separate 24 hour zone on the control box so that any damage to the foil due to careless window dressing, cleaning, or even deliberate tampering can be instantly recognized. For the highest security, inertia sensors should be used here and on other vulnerable windows, as these detect vibrations caused by prising and glass cutting, as well as breaking.

The front door will most likely be also the staff exit door in smaller premises. A magnetic sensor should be installed here, but connected to the exit loop which will allow a timed exit and entrance. An alternative is to connect a shunt key switch across the door sensor, or use it to switch on the whole system if the control panel allows for it. In this case the door sensor can be wired into the normal loop and so will give an immediate alarm if the door is forced open. If the addition of yet another key to the keyring is daunting, a lock-switch can be fitted which combines a switch with a lock.

If the staff exist door is at the side or rear, then this will be wired on the exit circuit or fitted with a remote control or shunt switch. The main shop door sensor will then be connected to the normal loop.

Where high security is required, other sensors can be installed in the shop or showroom. Pressure mats are inexpensive and can be placed in positions most likely to trap an intruder. The important thing with these is that chairs or displays are not inadvertently placed on them, as when under carpeting they can easily be forgotten. Being light, the chair may not immediately activate the mat, but could gradually sink into it and so trigger the alarm in the middle of the night. A classic false alarm situation! Alternatively, a PIR sensor wired to the alarm circuit can be used to protect a wide area and is less likely to generate false alarms. Do not forget the possibility of valuable stock spilling over into poorly protected areas during a seasonal rush. Make sure all areas likely to be used for storage are well protected.

The bell should be mounted high, at first storey level, but in a position of prominence at the front of the premises. Most systems have the bell mounted in a box with an anti-tamper switch, but consideration should be given to the point made in Chapter 6 as to the advantages of using a suitable open bell. It is louder, cannot be silenced by filling with foam, and is obviously not a dummy as many boxes are. In particular, the solenoid Tann Synchronome bells which are often supplied under other names, are well suited to open mounting and are virtually defeat-proof. If desired, a dummy bell box could be fitted some distance from the open bell as a decoy.

If the shop is in a rank with others that are also fitted with alarm systems it is not always easy to tell which bell is ringing, so a beacon lamp should be mounted alongside the bell. In the event of an alarm, a visible indication of which premises are affected will then be given.

A second bell should be fitted somewhere inside, near the back of the premises. This should scare off would-be intruders at the back who may not hear the front bell.

Panic buttons should be fitted at strategic points. Near the cash till is an obvious one, but one should be provided at any point where a member of the staff could be attacked, especially if the business is in the high-risk category. If this is so, thought should be given to having them at different parts of the building. One member of the staff could be helplessly isolated in the stores while a hold-up is taking place in the shop. An available panic button could transform the situation.

The control box should have a 24 hour panic-button loop separate from the 24 hour anti-tamper loop so that tampering and a personal attack can be distinguished. The box should also have an exit/entry circuit to provide a delay for the exit door and a power supply for running PIR detectors connected to the system. (PIR floodlights derive their power directly from the mains and are independent of the alarm system.)

For small shop premises, a two-zone control panel should be adequate. One zone will be used for the main detection loop, and the other for pressure pads and PIR detectors. Thus one protects the perimeter and the other the interior. With premises having living accommodation, one zone could be connected to the shop and the other to the accommodation. Thus the shop circuit would be 'on' in the evenings and weekends while the accommodation is 'off'.

With larger premises, four zones may be desirable, though not essential. Front and rear could be on different zones as well as having a separate space protection circuit. The fourth could be assigned to another area or perhaps the shop windows. If triggered, the indicators then make it immediately obvious which zone is responsible, which is very useful for investigating false and real alarms.

The type of main control for the panel depends on personal preference. The principal choice is between key operation and a key-pad of numbers like a calculator. The keypad would seem the best choice for those whose key rings are already too heavy, while a conventional key would be safest for those with a poor memory for numbers!

The panel should be concealed if at all possible. While control boxes are constructed from steel to resist a certain amount of force, they are unlikely to survive a frenzied onslaught with a sledgehammer or jemmy. Concealment is therefore the wisest course. There must though be sufficient time to get from the panel and through the exit door before the exit time elapses, so it cannot be fitted too far distant from the door.

One ingenious installation had a remote control switch included on the board of showroom light switches; it was identical to the others but fitted upside down. When all the switches were put up to switch off the lights at closing time, the alarm switch was thereby turned on. In the morning, the alarm system was turned off when all the light switches were put down. The switch appeared the same as all the others and anyone watching the staff enter or leave would have no inkling that an alarm system had been switched on or off. The only way an intruder could inadvertently switch the system off was by turning all the lights on which would be a very unlikely thing to do.

Of course alarms can be ignored and a quick snatch be made of valuables before making a hasty exit. One ingenious device which foils this was devised by a victim of many such events. The firm involved was the Winchester Garden Machinery in Hampshire and the device is a smoke generator. When the alarm is activated, the building quickly fills with thick white irritating smoke. Intruders cannot see their way about, the way out, or even their own hands. In an 18-month trial the device had been activated 15 times and successfully foiled all attempted burglaries.

The smoke is non-toxic and is actually used by the fire brigades and also the armed services for training purposes. It is available from Security Pyrotechnics, of Boston, Lincolnshire.

Another device is the Smart Water system. With this, the intruders are sprayed with a fine non-toxic spray which makes their clothing glow bright yellow under an ultraviolet light. It has a similar effect on the skin and is said to last up to six months after exposure to the spray. A number of high street stores have installed it including Marks & Spencer. Also known as Index Solution, the liquid was developed by Probe FX of Shropshire, and is manufactured by the Forensic Science Service.

While conventional alarms coupled with adequate physical security serves well enough for most situations, premises subject to repeated attacks could well be candidates for one of these systems. Although similar, their effects are different. With the smoke, the purpose is more preventive, confusing and disorientating the intruders, so they are unable to find what they are looking for. As a secondary effect, they may be so confused that they are still trying to find their way out when the police arrive, and so lead to their capture. The water spray may not prevent the removal of goods but will identify the culprits later. This assumes they will be found. The effect of the spray on stock is something to be considered.

Warehouses and stores

As before, the first step is to attend to physical security. Most warehouses consist of a large cavernous area having a cluster of several rooms attached which are used for offices and reception. Windows and doors of the offices need protection as described in the previous section. There may be little there of value to outsiders, but they could afford a means of access to the warehouse. Furthermore, the destruction of vital records by vandals could mean a greater loss to the firm than the theft of goods.

As with shop premises, the degree of security required will depend on the nature of the goods handled, though all goods have some attraction if access appears easy. So at least a minimum standard should be achieved.

There should be deadlocks on all outside doors and lockable catches on all windows. The condition of windows and window frames should be carefully examined. Many warehouses consist of old cinemas, converted chapels and the like. As no-one lives there, and a smart appearance is less important than with shop premises, woodwork and putty tends to rot quietly away unnoticed or ignored. This is less of a problem with modern buildings on industrial estates, but it should still be checked.

Coming to the warehouse itself, again all windows should be critically examined. As warehouses are usually draughty enough as it is, there is no need for opening any of these, so they should all be permanently sealed. For high-risk goods, grilles should be fitted to all windows. There is no need for the decorative scrollwork previously described, straight bars with intersecting tie rods are secure and practical.

The chief security hazard will probably be the loading bay doors. Often these contain a smaller access door. This is frequently flimsy with a simple springlatch lock. It would be good to consider whether this is really necessary, if it is not, a vulnerable point could be dispensed with.

Well-made stout wooden doors can withstand a considerable battering, but many loading bays have wooden doors that are insubstantial, and could be splintered away from their supports or shattered in pieces by a determined application of force. Ramming with a stolen car is not unknown. The thickness of the wood, and the number of cross members determine the strength of a door to a great extent. Also important is the frame supporting it and the nature and number of the hinges.

Wide doors often have two leaves that come together in the middle. As it is inconvenient to have a door post in the centre of the doorway, one leaf is usually secured by bolting it into the floor and at the top into the frame. The other leaf is then latched into the first. It follows that the security of the whole thing depends on those two bolts.

With large doors, it would not be too difficult to burst the bolts off the woodwork. Considerable improvement can be achieved if bolts are fitted to the other leaf at the opening edge top and bottom, and also a second pair on each door half way along (Figure 54). The bolts should be the largest obtainable and the screws should be long enough to almost penetrate the door. Dog bolts should also be fitted to both doors on the hinge side. This

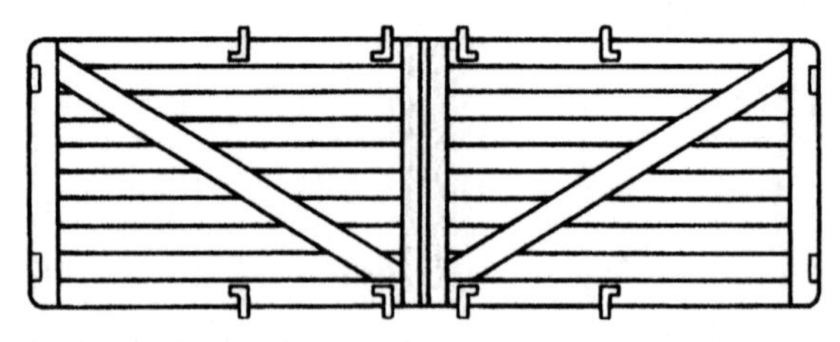

Figure 54 Double doors should have bolts top and bottom on both leaves as well as part way along.

should make the doors virtually burst-proof. A sloping concrete lip should be formed outside of the doors to cover the bottom gap. This prevents any attempt to saw through the bolts with a hacksaw and should also keep the rain out, as well as some of the draught.

Existing doors could be replaced by steel roller doors, and should be if the present ones are insecure. However, stout wooden doors well secured, take some getting through and can be more impact resistant than steel.

For very high security, a small wall of reinforced concrete about 18 in (450 mm) high could be built a few feet in front of the door. A lorry could back up to the wall and goods could be unloaded over it, but it would prevent a vehicle being used as a battering ram against the door. In cases where the floor of the warehouse is on a higher level than the ground outside, the loading door will also be higher, and the danger of ramming eliminated.

Warehouse alarm system

The alarm system should have at least two zones. One covering the offices and the other the warehouse. Office windows should be fitted with magnetic sensors connected to the loop, and the main door wired to the exit circuit. The door between the offices and the warehouse should be wired and also made physically secure.

For the warehouse, the loading doors should be protected by a magnetic sensor. If the doors are of wood, the sensor can go into the frame at the top in the usual way with the magnets at the top of the doors. Care should be taken to ensure that there is no 'give' there, otherwise wind or other forces may cause a false alarm. This could be done by mounting the sensor near a bolt, but not too near otherwise the metal may influence the magnetic field.

For metal roller doors the sensor shown in Figure 15 should be used. This is a low profile pod-shaped aluminium-encased switch that is fixed to the floor. If mounted at one side rather than in the centre, the wiring can be taken directly to the door frame; if fitted at the centre, the wiring would have to be run inside a tube buried in the concrete.

If the warehouse windows are sealed there is no point in fitting sensors to them. They could be protected with foil strip, but even this could be unnecessary if grilles are fitted. It could even be a source of false alarms. A window having grilles that was broken by a stone thrown by children, may not be a security hazard, but it could set off a full-scale alarm if the foil was broken. Much depends on the circumstances and the degree of security required, but the possibility should not be overlooked.

Actually it is not easy to give high-security perimeter protection to a large rambling building such as the average warehouse, especially in older premises. There is usually no ceiling, so roof entry could be very easy by just removing a few tiles. An alarm system should therefore include space protectors.

The choice and positioning of these depends on the size of the area and the layout and size of the storage units. It may not be possible to cover every square foot and inevitably there will be blind spots. The aim therefore should be to cover every section even though some parts of some sections may be blind. Intruders will not stay in one place, but move about to find what they seek, and then to make an exit. If the sensor disposition is well planned, it will be almost certain that one will be activated.

In most cases, PIR detectors will prove the most practical, but for very large warehouses an active infra-red or microwave radar system may prove to be more effective. PIR sensors cover up to 50 ft (15 m) with a fan-shaped detection area, or over 120 ft (40 m) with a pencil beam coverage. An overhead radar unit mounted 10 ft (3 m) high, can give a circular protection area of some 60 ft (18 m) in diameter beneath it, and larger coverage for higher mounting (see Figure 26). Soft materials such as plastic, cardboard and wood cast no radar shadows and so create no blindspots.

With lobe coverage, radar detectors have a range of up to 150 ft (45 m). These are Doppler devices, but beam-breaking radars extend to 500 ft (150 m). It should be noted that horizontally mounted Doppler radars are liable to false alarms due to penetration of the boundary walls and the sensing of movement beyond them, unless the sensitivity is very carefully adjusted.

Active infra-red systems use the beam-breaking method and the small models have a range of up to 40 ft (12 m) which is not as good as the PIR, but larger active units have a range of up to 1000 ft (300 m). Whatever their range, all beam breaking systems have a limited coverage as they only protect a narrow pencil of space between the transmitter and the receiver.

Generally, PIR detectors offer the best and most practical solution even if several have to be used. One mounted in each corner could protect a diagonal of some 100 ft (30 m), or a cluster on a raised post at the centre could have four aimed at a fairly high level toward the corners, and another couple aimed low to fill in the central areas missed by the others. This should give a coverage in excess of the 100 ft diagonal.

Goods yards, building sites

The combination of large area and the open air make it difficult to achieve high security. As with all other systems, much depends on the degree of security required. The first essential is a high, secure perimeter fence which for highest security should be continued for several feet below the surface of the ground to prevent excavation and tunnelling, and have outward projecting arms with barbed wire at the top.

Floodlighting is a valuable deterrent, but the high and increasing cost of electricity make economies worthwhile. One of these is to control the lighting by means of a daylight switch. This is better than a time switch because the latter needs continual adjustment throughout the year as the hours of daylight change with the seasons. The electricity costs will thus be considerably reduced by the switch during summer.

Another method of electricity-saving is to have a reduced amount of permanent floodlighting augmented by PIR controlled lighting. The permanent lighting would act as a deterrent, while any actual intrusion would bring on the extra lighting which would have an alarming effect on the intruder. Extra lighting could also be switched on when the main alarm system is triggered.

As the perimeter fence with its gate is the only means of entry, the alarm system is simplified by the use of a single type of sensor, the inertia switch. These need to be fitted at 10 ft (3 m) intervals around the fence and are controlled by an analyser which identifies the disturbances caused by climbing and eliminates those which could be created by random impacts. A lot of work is entailed in installing these as the wiring must be run in conduit tubing along the fence with joints to boxes at each sensor.

An alternative is to reflect an active infra-red beam around the inside of the fence (Figure 30). A 1000 ft (300 m) unit would cover 400 ft (120 m) allowing for the attenuation of three reflections. A 'double knock' analyser would be required to reduce the possibility of false alarms due to birds. The method is prone to adverse atmospheric conditions such as dew on the mirrors, and is triggered only when an intruder has gained access to the area. It must be installed several feet inside the fence to prevent an intruder jumping over the beam from the top of the fence, thereby wasting valuable site space. Although more expensive, the inertia detector method is better, especially where high security is required.

The control unit could have separate zones for each side of a large area, such as the grounds of a research establishment or a car compound where it would be advantageous to identify the location of attempted entry as soon as possible. For normal goods yards a single zone should suffice. The control box and analysers should be housed in a secure, on-site building.

As the site may be remote from habitation, a high-powered siren would be the most appropriate sounding device along with flashing beacons. For highest security, a telephone dialling device to inform responsible personnel would also be desirable.

Factory premises

Factories vary immensely in size, from the small workshop-plus-office to the large complex. In the case of the small ones the situation is similar to that of the warehouse and the same security arrangements can be employed. With the large complex, many different factors may apply to make each case unique, but we can make some general observations.

Perimeter protection is, as always, the first line of defence. High walls are a deterrent, especially if the top foot or so is painted with anticlimb paint. Pipes and ducting often afford ways of climbing on to flat roofs and from thence in through fanlights and windows. These too should be treated with anticlimb paint, but check that the temperature of the pipe does not exceed the maximum quoted for the paint which is usually around 212° F (100° C).

A problem with most factory complexes is that there is a large amount of space in and around the buildings, as well as storage yards, car parks and the like. These offer many opportunities for intruders to hide and evade detection, as well as commit acts of theft or vandalism.

It may not be possible to fully protect all outside areas, so the most vulnerable ones need to be concentrated on. Storage areas and car parks are a particular target. Floodlighting augmented with PIR controlled lights are a major deterrent, but such areas are best protected by closed-circuit television (CCTV).

A single camera with remote pan and zoom at one end of a car park can cover the whole area as no other sensor could. Other vulnerable areas can be similarly protected, and the output of all the cameras fed into a sequential switcher so that all areas are kept under constant surveillance. A motion detector can supplement the visual watch kept by the security staff:

Factory alarm system

If there are a number of separate buildings of any size, it is probably best for each to have its own independent alarm system with its individual sounder. A senior staff member working in that building would then be responsible to see that the system was turned on when the building was vacated. In the event of an intrusion, a separate system saves any confusion by clearly identifying the building concerned, and the local sounder fulfils the desirable function of scaring off the intruder before an entry is made.

Physical security of each building should be optimized as already described. Appropriate sensors can be fitted to windows and doors, and internal space protection detectors as required. Single-storey and flat-roofed buildings containing valuable equipment or stock are liable to be broken into through the roof. Protection can be afforded by fitting vibration detectors or acoustic sensors designed for wall protection (not those designed for glass as they respond to a different band of frequencies). (See end-of-book illustrations.)

Large complexes will undoubtedly have their own security staff who will have a central HQ. It may be desirable to have a link from each alarm system which, although independent, would signal to HQ that it had been triggered. This would give an immediate indication and identification of the intrusion point. It would also serve as a back-up for the sounder in case it had been silenced. The links could be by buried cable or infra-red beam which could also serve to protect intervening space (Figure 31).

It perhaps is stating the obvious to say that an outside telephone line should always be available to the security HQ, possibly bypassing the factory switchboard. Then emergency 999 calls can be made without any delay. Smaller complexes not having a 24 hour security guard could be equipped with a telephone dialler to call outside numbers in the event of an alarm.

Restricted access

There may be certain high-security locations that only authorized staff members are permitted to enter. One way of restricting entry is by means of an entry control system. This consists of an electrically controlled bolt which secures the entry door, and a card reader. The reader is a small steel case with a slot for inserting the card.

Cards have magnetic codes implanted on them, and when placed in the reader operate a switch which energizes and releases the bolt, allowing the door to be opened. When the door is closed the device resets. Unlike keys, cards cannot be duplicated, and should any be lost or fall into unauthorized hands, the reader can be altered to accept a new code. While restricting normal access, this system does not trigger an alarm if a forced entry is made.

Another method which can be integrated into the alarm system where very high security is required is to wire suitable sensors into the 24 hour anti-tamper circuit of one of the zones. Key-operated pass switches shunt the sensors to allow authorized staff to enter. Any forced entry triggers the anti-tamper alarm.

An interesting device which has much to commend it is the PushButton Combination Lock. This is a lock that has thirteen buttons and a door knob. It is entirely mechanical so has no need of wiring or batteries and is therefore not subject to electrical failure. The code is programmed by the user and can be changed at any time, useful when getting rid of a disgruntled employee! If desired it could be changed at regular intervals if many people had temporary access such as club members. A feature is the variable degree of security available. A code of up to thirteen digits can be used if required, which gives 8191 possible combinations. Most users settle for a four- to seven-digit code which gives a possible 5434 combinations. The knob has a built-in clutch which means that any attempt to force it simply makes it turn without unlatching the lock.

Several versions are available including a mortice type and a surface deadbolt type. The code digits can be entered in any order, but if an incorrect one is entered the lock will not release. If entered accidentally, a 'clear' button must be pressed and the code re-entered. The lock is obtainable from Quick and Slick, Andover SP11 6SZ.

The newly developing science of biometrics, which is the measurement and identification of biological characteristics, looks promising and is gradually becoming accepted. One method is the scanning of the iris of the eye. The pattern is unique to each person and so authorized staff can be recorded and checked. One system, the IriScan 2020, can examine an iris pattern and search a database of 10,000 records in about three seconds. This is being used in some cashpoint machines.

Another method is fingerprint and palmprint checking. FingerScan can check a print in half a second from a stored database. Voice recognition is yet another. A further possibility is the recognition of the vein pattern in the hand by infra-red cameras. This too is unique to each person.

The advantage of biometric systems is that they will respond only to the designated person. So they are not vulnerable to stolen keys or discovered code numbers. They therefore have a very high security factor, but there can be snags. Finger or palm print sensors need frequent cleaning, and voice recognition systems may not recognize someone who has a cold. All such systems need an associated computer to store the data, process the input and compare the sample. This means complication and expense, none of which is of any use if the system or the access door is vulnerable to physical attack. There can also be errors. Type 1 errors whereby authorized persons are rejected can approach 1 per cent. Type 2 errors, by which unauthorized persons are allowed access are much rarer, but they may occur due to a system fault. Simpler systems are often the best – which is why the PushButton lock with regular code changes is probably as good as anything.

Public halls

Church halls, community centres, village halls and the like, generally have few if any valuables, but are vulnerable to youths who think there may be some cash or cigarettes on the premises, and also to vandals. High security is thus not needed, but all windows and doors should be physically secured as described previously. It is worthwhile installing an alarm system, but if there are a large number of windows it may be more practical to opt for a couple of PIRs. Any doors and windows that are obscured from public view should be wired.

The main problem is likely to be due to cleaners, hirers and others forgetting to switch the system on when leaving. The ideal set-up is to have the control box out of the way, and the switching accomplished by a switch-lock on the main entrance.

Control boxes designed for the purpose have facilities for a buzzer to be fitted near the entrance. If a sensor is activated by a window left open when the key is turned, the buzzer sounds continuously. Then, pressing a cancel button on the control box resets the system. If all is well the buzzer sounds for a few seconds and stops.

Strangely, in view of the obvious demand for such a unit, they are not easy to come by and some compromise may be necessary. This most likely will mean no warning buzzer, so the alarm will sound if the door is locked with a window switch or other sensor active. As there is also no means of telling whether the system is working or not, a regular test would have to be made by the caretaker or other responsible person.

A bell should be fitted high on the front of the building if, as most are, it is a single storey structure. Another should be fitted high inside the main hall.

Hotels

A hotel can be one of the worst headaches from the security angle. It can be protected just as a private house during the off-season period, but when the season is in full swing, there are guests wandering in and out at all hours and keys are handed out to all. There are likely to be valuables left in guest's rooms, although a prudent hotel-keeper will warn against this. Really, a hotel can be a burglar's paradise.

Any conventional alarm system is likely to come unstuck with guests leaving bedroom windows open and even perhaps opening lounge windows. The main danger is that of an intruder slipping in during the day, while guests are out, or perhaps when at dinner and the staff are occupied. Most of the bedroom locks fitted in the majority of hotels are child's play to an experienced burglar, while the use of a ladder and window-cleaner's outfit could easily disguise an entry by a window.

One system which can be suggested, though unconventional, is a practical method of protecting guest's belongings in the bedrooms. All bedroom locks could be changed for lock-switches which could only be locked from the outside. Internal bolts could be provided for security of guests at night. The lock switch could be connected in series with one or two pressure mats near the window and dressing table, the whole being wired into a central alarm system along with all the other bedrooms. Each floor should be on a different zone so that an alarm could be quickly localized.

When the door is locked, which only can happen when the guest is out, the pressure mats are on guard to trap anyone entering through the window or forcing the door without unlocking it. When the guest returns, the door is unlocked and the mats de-activated. As the door cannot be locked from the inside, visitors cannot trigger the alarm inadvertently. Indeed there is no need for them to even know that an alarm system is operating in their bedroom, although some proprietors may choose to inform guests as an example of the management's concern for their welfare.

It may be considered desirable to have a buzzer operated in the office or staff section rather than an alarm bell which could disturb and cause consternation to other guests, who may mistake it for a fire alarm. A discreet investigation can then be carried out without upsetting anyone else.

The system would need to be on 24 hours a day and so could be a separate one from the main system having external sounders, which would be used off season. Alternatively, a control panel could be chosen which had 24 hour circuits that sounded a buzzer, as well as the normal switched detection loops that operated external sounders. Then the single unit would control everything.

In addition to intruders from outside there is the possibility of guests straying into private areas such as behind the bar. No doubt this would be locked when not in use, but pressure mats at strategic positions could serve as additional protection. These would need a separate system as it would have to be switched off at appropriate times; a very simple latching battery operated buzzer circuit should suffice for this.

While most guests pay by cheque, there no doubt will be large amounts of cash on the premises especially at week-ends. Steps should be taken to ensure that these are adequately protected.

10 Domestic systems

Much of what has appeared in the previous chapter, applies to domestic systems, though on a smaller scale. However, as private home installations tend to be a separate part of the installation scene, we will here repeat for convenient reference, the relevant points with appropriate modifications and observations.

First, the various items discussed in Chapter 8 on general security should be carefully reviewed. Nowhere, is physical security lacking, more than in the average home. The front door may indeed sport expensive security locks and a bell box hang ostentatiously on the wall, yet all the while the rear windows and back door are almost begging to be broken into. As the principal means of entry, these should be made quite secure before there is any thought of installing an alarm system.

After breaking in, the burglar's first task is to establish an exit route by opening an exterior door. This enables him to carry out bulky items and also to quickly escape should he be disturbed. Usually, the one chosen is the back door, and often he will bolt the front door to prevent a returning occupant entering and discovering him.

Bolts on the front door are therefore of dubious value, they cannot be used when the premises are vacated, only at night, yet they can well serve the purposes of the intruder. Rack bolts (those that are lockable) fitted to back and side doors prevent them being opened from the inside, as well as adding to the resistance of the door to being forced from the outside. The front door can be secured against inside opening without a key, by fitting it with a security deadlock or a double-locked deadlatch. However, the key should be left in the lock or the deadlatch not double locked at night.

Windows, especially at the rear where they are less likely to be observed, are the most usual means of entry. Intruders rarely if ever enter through broken glass, it is too risky; they prefer to open the window after forcing, removing or breaking it. A locked window effectively thwarts this. All new windows should have integral locks especially on the ground floor, while non-locking windows can have them fitted. Puttywork which is old and cracked should be replaced as otherwise it can be broken away and the glass pane lifted out. DIY aluminium windows popular a few years ago that had fixing screws accessible to the outside, should have the screws glued, or the window replaced. All frames should be examined for weakness.

Having seen to these basic elements of physical security, further measures and the type and extent of the alarm system, should be carefully weighed against what is really necessary. The standard of security elsewhere need not be as high as it is for business premises, unless in a high-risk area or having a bad history of previous burglaries. Intruders are usually opportunist thieves, often teenagers, looking for a quick and easy 'job'. Sufficient protection, both physical and electronic, to thwart such individuals and make them look elsewhere, is all that is required.

Unnecessary security can not only be a waste of money, but what is even worse, can impose irksome restrictions and limitations on everyday life. The occupants could feel that they are living in a fortress. Remember that the security system must be lived with, month in, month out, year in and year out, for a considerable time to come. The goal should therefore be *'friendly security'*, friendly that is to the occupants. Security should cause the least inconvenience, although some must be accepted. The danger with inconvenient over-security is that sooner or later it is relaxed and neglected. The important thing is to identify and protect the vulnerable points while being less fussy over low-risk areas. Also if there is more than one way of achieving a similar degree of security, chose the one that will be the least inconvenient to operate and use.

For example, the chances of anyone tampering with the sensor wiring is slim because people having access to the house are usually only family members or friends. A 24 hour anti-tamper loop is thus not really necessary.

The timed-exit routine can become very irksome; having to get out and shut the front door before the alarm goes off, then facing the same procedure

Figure 55 Glass-panelled front doors are extremely vulnerable especially if ordinary locks are fitted. Such doors should always be wired to the alarm. Solid wooden doors with security locks need not be wired as these are not likely to be attacked. All back doors into the main building should be wired.

again when entering, perhaps with armfuls of groceries. If the front door and frame is substantial, and fitted with a security lock, it is unlikely to be attacked. So, it need not be wired at all and the exit/entry hassle thereby completely abolished.

As a precaution though, a pressure mat could be put just inside the door in a position that would be stepped on by anyone not knowing of its presence, but easily avoided by the occupants who do.

If, however, the door has glass panels and there is no intention of fitting a more secure replacement, a door contact should be installed. But then it could be shunted by a switch in the deadlock, or the lock-switch used to remote set the control panel if it has that facility (see Figure 55). Thus the exit is protected with no inconvenience.

Windows

Permanently sealing some windows has been advocated for business premises but it would undoubtedly bring protests from the family during hot weather. Window security can be maintained by the fitting and using of window locks as previously described. Those rarely opened can be kept locked, while the key for others can be kept nearby to be used as required. A little inconvenience may thereby be caused, but it is not too great, and window locking is an essential part of the security.

The point made in Chapter 8 regarding fire is worth repeating. An easy escape route from first-floor windows is essential even at the loss of some security. First-floor front windows are rarely, if ever, entered by intruders because of the difficulty of access and the risk of being observed. Yet these are usually the easiest to effect a rescue from fire-escape ladders. So they should never be permanently sealed but be readily openable from the inside. It is better to be burgled than burnt!! This is why front doors should be easy to open at night and the key kept in the lock.

All ground-floor windows should be wired to the system, as most entries are made through them. Even first-floor windows are vulnerable if there is possible access from a porch, flat roof, drain-pipe or bay window (Figure 56). These should also be wired, and all windows, even fly windows should be kept shut and locked when absent even for a short while. A ten-minute trip to the local shop is often all the watching intruder needs.

Large areas of fixed glass may tempt an intruder to cut a hole large enough to enter. Window foil strip is one solution which has the advantage of being seen, so serving as a deterrent and preventing damage to an expensive pane of glass. Especially vulnerable locations could be protected by scrollwork grilles which can enhance appearance as well as increase security (Figure 53).

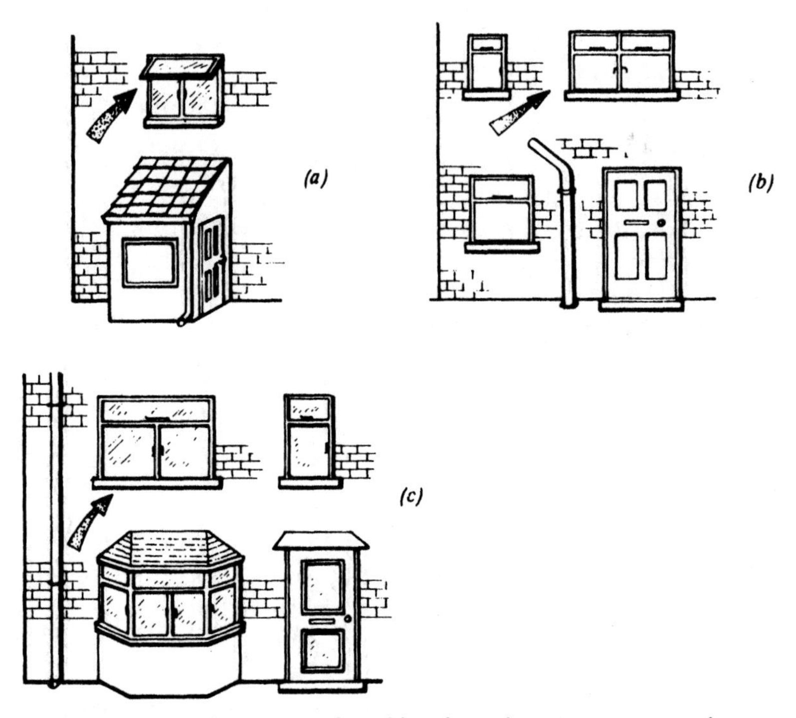

Figure 56 Upstairs windows are vulnerable where there is easy access from: (a) porch or lean-to; (b) drain pipe; (c) bay window, or any other nearby support. These should be wired, and kept closed when not at home.

Another problem arises with bays, which can have up to eight opening windows, all needing sensors. An alternative is to place a pressure mat in the bay where it is certain to be stepped on (Figure 57). A second line of defence can be achieved by wiring the room door with a contact, as an intruder is sure to open it to explore elsewhere and also establish an escape route.

Figure 57 Multiplicity of opening windows in a bay may prove a problem. A pressure mat under the carpet may be a more practical solution. To make sure, wire the room door as well.

Exterior doors

All rear and side doors should be wired with sensors, as well as being made physically secure by fitting with deadlocks and rack bolts. This applies to doors that actually give access to the main building. However it is pointless to protect conservatory doors in this way or fit them with sensors, because these structures are so vulnerable. They have plenty of windows, and glass or plastic roofing, as well as a flimsy door in most cases. A locked conservatory door would pose few problems to a burglar, there are just too many other ways past it.

Instead, concentrate on the door that leads from the conservatory into the house itself and treat it as an exterior door (Figure 58). There is little of value in the conservatory, the intruder merely wants to get through it into the main building where the valuables are.

Figure 58 It is hopeless to try to protect a conservatory, there is too much glass, corrugated plastic roofing and a flimsy door. It is better to concentrate on the conservatory door into the house (b). Fit a sensor and keep the door locked. The same may apply to many built-out kitchens.

The same thing applies to some kitchens. Many of these are single-storey structures built on to the main premises. Door, windows and even the roof could afford a means of entry. So although the outside door should be locked or rack-bolted, the kitchen door into the rest of the house should have an alarm contact and preferably be bolted from the house side, especially during prolonged absences such as holidays.

Roofs

Although it is unusual for an entry to be made through a roof, it is by no means unknown with a single-storey building such as a kitchen or home extension. It is certainly not difficult, although it may take a little more time than the usual methods. Tiled roofs are especially easy, requiring only the removal of a few tiles, tearing away any underfelt, and kicking a hole in the ceiling beneath. The burglar has no qualms as to the damage and mess he causes.

It follows that bungalows are especially vulnerable and many have been broken into in this manner. In this case, breaking through the ceiling could be hazardous as being unfamiliar with the layout, the intruder cannot tell where he is likely to land. So, the loft trapdoor is virtually his only means of entry from the roof cavity. Bolts fitted to the underside put a simple and effective stop to this.

The establishment of an exit route is particularly vital to him if he did succeed in getting in this way, as an exit back through the roof would be very difficult. So the fitting of rack bolts and sensors to all exterior doors is especially important with bungalows.

Interior sensors

While the protection of windows and exterior doors in the manner described goes a long way to establish security, it is always wise to have some interior sensors as a second line of defence. They add little to the cost, create no inconvenience, and so are well worth it for the extra security they afford.

It could be that an entry is made at some unprotected point such as through the roof. But the alarm could still be triggered by an interior sensor if it is in the right position.

Magnetic door contacts should be fitted to rooms where valuables are likely to be found, such as the living room and the main bedroom. In addition, pressure mats are a convenient trap for the intruder, and should be used under carpeting in the living room near the video and TV, and in front of cabinets containing such items as a trophy collection. In the bedroom, a pressure mat can be placed in front of the dressing table as this is a certain stop for a burglar looking for jewellery.

Under the stair carpet near the bottom on the second or third stair, is another good place for a pressure mat; the first one is often stepped over. If the front door is not wired, a mat in the hall is essential.

Passive infra-red (PIR) detectors may be used in large living rooms with several windows, french windows and doors, where it may be difficult to protect all with magnetic switch sensors. The loft area of a bungalow is another place where the PIR could prove its worth. In short, any large area that is vulnerable due to multiple possible entry points or having especially valuable items would benefit from having a PIR installed.

In addition to applications inside the home, the PIR can be used to great effect in increasing security outside most homes. The PIR controlled floodlight can be adjusted to stay on after being activated from a few seconds up to 15 minutes. It can be fitted over a garage, on the rear wall, over the side path of semi-detached premises, or anywhere where intruders may lurk at night. The light can often be seen to come on from indoors, thereby warning of approaching visitors or prowling intruders.

It also has the advantage of conveniently providing light when required for the occupants or legitimate visitors when approaching the house. Even if activated by an animal, no harm is done as there is no connection to the alarm system and so no false alarm is generated.

Panic buttons

The possibility of attack while at home by a bogus caller at the front door, or an intruder breaking in or walking in at the back, while the door is unlocked, is unfortunately ever present. Panic buttons connected to the alarm system can sound the alarm even though it is not switched on, and so are ever ready for an emergency. Once triggered by a panic button, the alarm continues to sound until reset at the control box and the button is released by means of a key.

Any number of panic buttons can be installed and some thought should be given to their location. An obvious place for one is near the front door so that it can be actuated if a caller tries an attack. However, the attack may push the occupant backward, and if the button is right by the door, it could then be out of reach. It may therefore be an advantage to place the button a little way back from the door.

It should not be installed in such a position where it could be accidentally set off, yet not placed so that it would be difficult to find in an emergency. It should be so placed that the hand would go to it without needing to look. Some experimenting should be tried with the button held in different places to determine the optimum. It should be out of reach of young children who may think it great fun to set off the alarm, but not beyond the reach of older children who may have to use it. Adult shoulder height would seem to be about right.

Another location requiring a button, is the main bedroom, perhaps by the bedside. This could be used if an occupant happened to be upstairs when an intruder broke in on the ground floor. If near the bedside, it could serve if someone was resting during the day or ill in bed when an intrusion occurred.

The house layout will determine which if any other positions may be required. Some possibilities are: in the kitchen, by a home-worker's desk, or anywhere where the occupant could be isolated from the principal buttons by an intruder forcing an entry.

One general point regarding panic buttons for domestic use could be made here. Most operate in conjunction with a 24 hour closed loop and so the contacts are normally closed. They tend to be over-engineered, large, ungainly, and need a key to reset. These features are undoubtedly of value for maximum security in business premises where there is a possibility of previous tampering as a prelude to a planned attack. But this is not the case with a domestic situation.

In the home, most buttons are unattractive and conspicuous, and can spoil the look of any decor. It is understandable why some householders are reluctant to have them appearing here, there and everywhere. Yet if the control panel offered a latching normally-open panic circuit, there would be no need to use a special obtrusive button that requires resetting, and an ordinary bell push would do the job. The push could be inconspicuously mounted anywhere, such as on the side of a door or window architrave, and there could be little objection to its appearance.

Though security would be slightly compromised, the increased risk in a domestic system would be negligible, in fact it would likely be reduced as the householder may agree to having more than the minimum number of buttons. This then is another example of 'friendly security'. Unfortunately, the latching normally-open panic circuit is a rarity among control panels.

Control and zoning

The control panel needed for domestic installation is similar to that for business premises, but simpler. The anti-tamper loop is not required, neither is a timed exit circuit if any of the alternative suggestions are to be used, but most panels have it anyway. The basic requirements are: detection loops, panic circuit and pressure mat circuit. With some panels the mats are connected across the detection loop and anti-tamper circuits so the latter would be required even if not used as such. Sometimes the panic buttons are connected to the anti-tamper circuit instead of having their own separate circuit. An auxiliary switched supply will be required if PIR detectors are to be used.

Many home system control panels have just a single zone which may seem sufficient, but there is an advantage in having a two-zone or even larger system.

Many burglaries have been committed by the intruder entering at the back of the house and slipping upstairs to take jewellery from the bedroom while the family watched TV at the front. Having the back of the house on a separate zone which could be switched on when the family is engaged at the front, thus makes good sense.

Another use for zoning could be the wiring of all upstairs sensors on a separate zone. These would be off at night but switched on when the house is empty during the day along with all other zones. Either of these options could be entertained with a two-zone panel, but if both were desired, a four-zone unit would be necessary. The fourth zone could then be used for the garage.

The control box should be located within the protected area so that it cannot be reached without triggering a sensor. Yet it must be possible to leave the panel after switching it on, without setting off the alarm. Normally this is done by making use of the timed exit facility, but as we have seen, a more convenient alternative is to have a lock-switch on the exit door that either shunts the exit door sensor or remotely switches the system. A really stout door can dispense with contacts altogether, protection being afforded by a strategically placed pressure mat in the hall.

Control panels are often fitted in the hall, but there is no reason why they should not be located elsewhere. The important thing is that they should be concealed, a few hefty blows with a crowbar or jemmy could soon put paid to it and silence the alarm. It is not always easy to hide the box in a hallway; but it could be put in a cupboard behind some carefully placed 'junk', or on a wall behind a clothes rack.

Much depends on the layout. A too remote and secure location may create inconvenience when leaving and entering the premises, and result in not bothering to switch on when popping out for a short while. This can be an expensive mistake, as many householders have discovered. So, as long as it is not actually conspicuous and inviting attack, intruders are unlikely to go searching for it with the alarm sounding.

An important point too, is the power supply. Often with DIY installations the box is plugged in to an ordinary electrical outlet socket on the end of a long lead. This creates the possibilities of the plug being pulled out to plug something else in, or the socket being switched off. Although there may be standby batteries these will not last for long if the power remains disconnected for any length of time.

A special fused wall connector should be installed for the purpose which prevents inadvertent disconnection. This should be done by a competent electrician.

Alarm bells

The main bell should be mounted on the exterior of the building in a position that is inaccessible except by a ladder. This means high up, away from drainpipes, bay windows or other means which could afford access. The bell can be fixed vertically on the wall (most units are fitted in this position) or horizontally under the eaves if room permits. The latter position

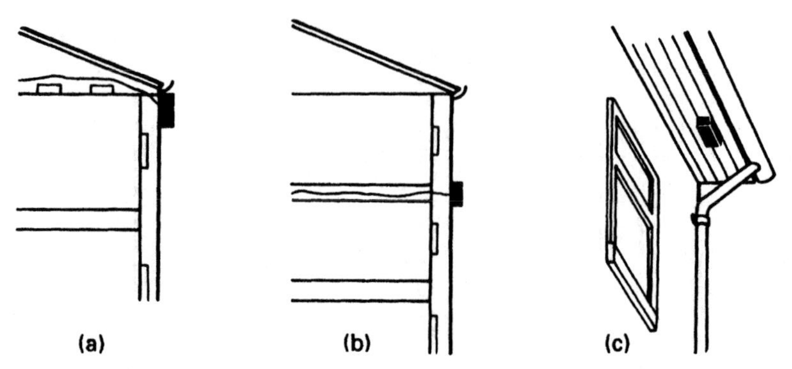

Figure 59 An outside bell can be fitted above room ceiling level then wiring can be run above ceiling to be dropped to where control box is located (a). If passing through an intervening storey it may be most convenient to take it down a corner or door-frame. With three-storey buildings (b) it is best to mount it between first and second floors. This is high enough to be beyond access, any higher would reduce sound volume at ground level. A good place is under the eaves (c) if they are wide enough. Fitting and drilling is easier, and wood can act as a sounding board.

is easier to install as it means screwing into wood instead of having to drill and plug a brick or stone wall. Also the wiring can be taken easily through the wood into the back of the bell, and the wood acts as a sounding board, so increasing the sound volume (Figure 59(c)).

Wall mounting means making a hole right through from the inside, and this will govern the position, as it is undesirable for the wiring to come out in the middle of a room wall. It is best for it to pass through the wall between the ceiling and floor above it, so that the cable can run under the floor and drop down through the ceiling immediately above the control box (Figure 59 (a) and (b)).

The important thing is that the bell should be in a position where it will be both seen and heard. This usually means fitting it on the front of the house which is nearest the road, but not necessarily. The road may be little used, perhaps a cul-de-sac, yet there may be houses nearby, facing the side or rear of the premises. As there is little point in putting the bell on the side of a building furthest from those likely to hear it, it would be better to put the bell on the facing side.

It should also be readily seen when approaching the premises and thus serve as a deterrent. If there is a conflict between these requirements, two external bells should be used, each on a different side of the building. In such cases it is essential that low-consumption units be used, otherwise the duration that standby batteries could sustain an alarm would be curtailed.

Whether or not two external bells are used, it is a good practice to fit an internal one. This would rouse sleeping occupants and scare off intruders who may have penetrated the perimeter defences and triggered an internal sensor. The bell should be mounted where it is difficult to reach and can be clearly heard over the whole house. A good position is at the top of the stair well, which satisfies both these requirements.

Door entry systems

Although not part of an alarm system, remote door entry devices are of value to the elderly, infirm, flat dwellers and others. The system comprises a loudspeaker/microphone unit that is fitted to the door, a telephone to which it is connected inside the premises, a door lock-release and a power unit.

The speaker and telephone units allow the occupant to converse with the caller. If satisfied as to his identity, the occupant presses a button and the lock-release disengages the door catch, allowing entry.

The lock release is fitted to the door frame in place of the normal staple into which a springlatch or deadlatch engages. The existing lock thus continues in use. As the caller must pull the door behind him to secure it when he leaves, the system cannot be used with a security deadlock, but it can be used with a deadlatch.

Two separate power supplies, one d.c. for the speakerphone, and the other a.c. for operating the release are contained in the power unit. Five wires are needed so six-core cable is required.

Two or more phones can be connected to the one door unit, and multiple systems are commonly found in high-rise flats. The caller selects the flat number he wants by entering it on a keypad which is part of the speaker unit. Some of these systems have a time switch which releases the lock when a tradesman's button is pressed during restricted periods. At other times the button is inoperative.

11 What not to do

The previous chapters have described the essential features of intruder alarm systems as well as practical general measures to prevent unwelcome visits from the light-fingered fraternity. It should enable the reader to assess any proposed system and distinguish good advice from bad. This chapter is by way of a test. It describes a true-life situation in which the author was involved, in which a good existing intruder alarm system was replaced with another on the advice of an 'expert' professional installer. The new system violated nearly all the rules and principles that have been described. In reading this account, see how many errors you can detect. The answers will be given at the end of the chapter, along with references in the text to where the points are dealt with.

The premises

The building was a church hall 15 × 9.6 m, in which regular services and meetings were held. There was an annex, mother's room, library, toilets and a foyer. Some years ago the building had been broken into on several occasions, always from the rear of the premises which was hidden by a high dual-carriageway embankment. The front faced onto a cul-de-sac, with terraced houses nearly opposite. The rear was thus the obvious place for a break-in, and none had ever been attempted from the front. Nothing of value was kept on the premises; being in an inner-city location, the miscreants were not professional burglars, just local youths hoping to find something worth taking.

The original alarm system

An alarm system was installed which consisted of a detection loop with magnetic contacts on all accessible windows and doors. The loop was a single wire; as it was buried in the plaster for most of its length and members of the public had no access other than at scheduled meeting times. The wiring was not vulnerable to tampering and so a 24-hour loop was not considered necessary.

A large 8-inch Tann Synchronome bell was fitted to the exterior, with a smaller 6-inch bell inside the main hall. Setting and unsetting was by means of a lock-switch in the main entrance. The whole system was switched on by just locking the door, and switched off when unlocked. A buzzer near the door sounded and died away to indicate the alarm was set and all was in order. If the buzzer sounded continuously, this indicated that a door or window had been left open, and would have to be corrected before another attempt was made to secure the premises. Thus the setting was virtually foolproof.

The two bells could be tested silently by a single operation at the control box, this being done at regular intervals. Less frequently an actual sounding test was made. Provision was made in the control unit for the addition of PIR detectors if required, also a mains floodlight to come on when the alarm was triggered, although these were not thought necessary and so not fitted.

The control unit was a Sureguard which had a very high reliability factor, as by an ingenious design, all of its many functions were achieved with the use of a single transistor. In the event of a power failure, the system could automatically switch over to a battery back-up, the power consumption and battery capacity being such that it could run the system continuously for over a year.

Among the general security measures taken were a decorative iron grille fitted inside a fanlight in the ceiling of a passageway, and hinge bolts fitted to three outward opening firedoors at the rear of the premises.

During the ten years or so that the system was installed, it foiled at least two known attempted burglaries, and sight of the formidable-looking bells and occasional test soundings no doubt deterred many more.

Of course it did not prevent vandalism to the outside of the premises, and on one occasion serious damage was done, apparently by a maniac driving a JCB to the perimeter walls and entrance gate. This spurred the building committee to review security arrangements.

Their first priority was to secure the outside boundaries. This was done by high spiked steel fencing at the rear, high thick bushes planted around the frontage, and high iron gates at the entrances. These measures made unwelcome access to the grounds and car park virtually impossible.

The new system

Next they decided to 'upgrade' the alarm system to a 'more modern' one on the advice of an installer, but against the author's advice. The existing system was to be stripped out and a proposed new system fitted. A description of this is as follows:

The new system was to be set by a keypad near the main door.

Detection was to be by means of dual technology detectors consisting of a PIR sensor and microwave device in the same case. For the main hall a single 12-metre-range detector was proposed to be mounted in one corner aimed diagonally across it, the rear fire doors being on the wall furthest from the detector, a small-range one in the passage leading to the toilets, mother's room and library, one similar in the mother's room, and in the library, and one in the annex, making five in all.

The existing detection loop with window and door contacts were to be disused and disconnected.

For the sounders, the Tann Synchronome bells were to be removed and discarded, although the outside one had been recently replaced and was brand new. In their place a siren was to be fitted to the front outside wall, and another in a weatherproof case inside the hall in the same location where the bell had been. In addition, dummy sirens were to be fitted to the other three exterior walls.

The Sureguard control unit was scrapped and an ordinary unit fitted in its place. This was in a concealed and not readily accessible position. It had numerous zones, so each PIR detector was connected to a different zone to enable the triggered one to be identified from the control unit if a break-in occurred. There was no zoning on the original Sureguard.

A small dummy video camera was installed in a dark corner of the foyer aimed at the main doors which are glazed with reinforced glass. However, the camera cannot easily be seen from outside in the daytime because of its position, nor at night in spite of its tiny red LED because of reflections in the glass from street lighting.

So there was the proposed system. As a result of vigorous objections, modifications were made. One was against the dispensing with the perimeter loop. As a result a loop was added, but the original wiring was not used; instead wiring with an anti-tamper loop was run and connected to the original magnetic switches. The system was approved by the committee and installed.

The hinge bolts fitted to the fire doors had been removed and never refitted when the doors had been previously replaced. No replacements were planned or fitted during the security overhaul.

Now, read the description again carefully, and try to recall the various points made in previous chapters. See how many faults you can find with the new system and its modification before you read the following pages for the answers.

The faults

1 **Security level incompatible with risk factor.** The committee had rightly given first attention to the boundaries of the property, installing fences, bushes and gates that were virtually impenetrable. Professional thieves after real valuables would no doubt find a way in, but inner-city vandals would be deterred and look elsewhere. The break-in risk factor had thereby been greatly diminished to almost zero.

The principle that should be followed is that the security system should be matched to the risk factor. A high risk requires high security, whereas low risk requires less. Bank vaults need more security than a private house. Excessive security is not only a waste of money, but can cause continuing inconvenience to those who have to live with it (see pages 4, 110 and 128). The object in this case was to 'improve' the alarm system (although that was not the result). Yet this was quite unnecessary; the existing system had proved itself and was more than adequate with the reduced risk. Money was wasted that could have been spent more usefully.

2 **Loop versus volumetric detection.** The original plan proposed for the new system abolished the loop detector in favour of volumetric detectors. However good these may be, they still allow an intruder to enter. Vandals could do a lot of damage (thowing paint or even a petrol bomb – it has happened!) in the few seconds they are inside before an alarm sounds. It is infinitely better to keep them out! There are certain situations where a loop is impracticable and volumetric detectors are the next best thing. Sometimes cost may be a prime consideration, and a few PIRs may be cheaper and easier to install than running a loop to every window and door. But in this case a loop was already there! There was no excuse for this serious violation of such a basic and important principle. The fact that a loop was finally added was no credit to the installers.

3 **PIR range.** Proof of the folly of relying entirely on volumetric detectors is shown by the range of the one in the main hall compared to the area it had to cover. Did you compare the range of the main hall detector with the dimensions of the hall? The specified range was 12 m, and the length of the hall 15 m. The diagonal from the detector to the furthest corner is 17.7 m. The furthest wall has two fire doors which are at the rear and have been attacked in previous attempts at entry. These vulnerable doors are therefore out of the specified range and unprotected.

Actually, the quoted range is somewhat nominal and the actual range is usually greater; however it is unwise to rely on this. This was shown by a walk test. Detectors have an indicator light that illuminates without sounding the alarm when movement triggers the device, thus enabling the coverage of the device to be checked. In this case the walk test revealed that the range falls about 3–4 m short of the furthest fire door which is one of the most vulnerable parts of the building.

A similar error was found in the annex. The small-range device used there does not reach to the annex fire door but falls short by about a meter. The installers should have checked these coverages.

4 **Physical security neglected.** As previously noted, the hinge bolts on these vulnerable doors were not replaced. This, combined with the lack of coverage by volumetric detectors, makes the doors even more vulnerable, especially as they were the objects of attack on previous occasions. This betrays a common lack of understanding of basic security needs. Physical security always comes first and is of more importance than any alarm system (see page 98). It was only as a result of vigorous objection that a loop was installed which does now offer the needed protection.

5 **Unnecessary detectors.** Seeing that a loop has been installed, the volumetric devices are redundant. Anyone attempting to break in would trigger the loop before getting anywhere near the detection zone of the PIRs. It makes sense to have both in high-risk situations where determined thieves may use tunnelling or roof removal and in such cases the volumetric devices would serve as a back-up. This is certainly not the case here. So the cost of five detectors could have been saved.

6 **High operating current.** One disadvantage of PIRs is that a relatively high current is needed to keep them on standby compared to a loop. Dual-technology devices need even more. Five therefore consume an appreciable amount. This is no problem when operating from the mains, but if switched over to standby batteries, they would last a matter of hours rather than months, as was the case with the Sureguard loop which had an extremely low circulating current.

7 **Unnecessary anti-tamper loop.** Anti-tamper loops are essential where there is unsupervised public access to buildings and where the loop or its components could be tampered with. Where there is no such possibility they are not needed (see pages 11, 22). Some installers may run an anti-tamper loop in the same cable as a matter of routine with a new system. In this case it was unnecessary, but as existing wiring was in place the rewiring was not needed at all.

8 **Unnecessary zoning.** In large premises such as factories, department stores, or flats, zoning that identifies which part of the circuit has been triggered is very useful as it saves time in making unnecessary searches. Sometimes in smaller installations, a two-zone system can be advantageous (see page 14). For small premises such as this, six zones are quite over the top. Any attempted break-in would be immediately apparent by a forced rear door or broken window. One would not have to go the

control box which is concealed and not quickly accessible, to discover the obvious. No disadvantages were experienced with the previous single-zone system.

9 **Bells versus sirens.** This really was an act of vandalism; the superb Tann Synchronome bells were discarded for sirens housed in nondescript boxes. Sirens for buildings may be advisable where the premises are situated some way from any road or other building, as they may stand a better chance of being heard. Otherwise, as in this case, with a terrace of houses just across the road, the bell is to be much preferred (page 81).

   Sirens are associated with car alarms, and these go off with so little provocation that no-one takes any notice of them. A bell is definitely associated with buildings, and although false alarms do happen, they are not so frequent as false alarms from cars. So, a sounding bell is much more likely to attract attention than a siren, and the stridency and volume of the Tann Synchronome, being a lot better than most, is more than adequate. As these bells are exposed, they cannot be confused with dummy boxes which are getting rather too popular (see page 76).

10 **Reliability factor.** Ordinary control units achieve their various functions with dedicated integrated circuits, or chips as they are called, that are designed for the purpose. These contain a large number of individual transistor elements. Generally, reliability is good, but every transistor or other electronic component has an average failure rate. So statistically, the more there are, the greater the probable failure rate and the lower the reliability factor.

   The Sureguard control unit was specifically designed to counter this and achieved a high reliability factor by using just a single transistor to perform all the functions it offered. It also permitted silent simultaneous testing of both bells as well as occasional actual test soundings. All this was sacrificed by the substitution of a run-of-the-mill unit.

11 **The dummy video camera.** These can be quite effective as a deterrent. However, referring back to point 1, as the risk factor was considerably reduced, there was little need for it. Even so, it was fitted in the wrong place, where it could not be readily seen, and at the front of the premises where no break-in had ever been attempted. It would have been more effective to have mounted it outside, at the rear, to apparently cover the vulnerable rear fire doors.

12 **Exit and entry pad.** The proposed setting and unsetting was to be by a keypad to replace the door lock switch previously used. With a number of people entering and leaving the premises for cleaning, committee meetings, etc., it is almost certain that, sometimes, setting would be forgotten and the premises left unguarded. Here again strenuous

objections overcame this folly, and door-lock setting was fitted with a nearby bleeper to indicate it – just as it was before.

Conclusion

Well, how many did you spot? If only one it was one more than the 'experts' who planned and installed it, but hopefully as an attentive reader you did better than that. Other minor details could be mentioned, such as the use of a siren inside the hall and a weatherproof box for it which is far less aesthetically pleasing than the attractive Tann Synchronome bell it replaced. The cost of the new system was £300, money which could have been better spent on other things that were needed.

There are many lessons to be learned from this sad but true tale. First there are the actual technical points listed above. These provide a good lesson in the sort of errors likely to be encountered in proposed systems, and so what to watch out for. These are:

- adjust your security level to the risk factor;
- use PIRs only as a backup, but if a perimeter loop is impossible, make sure the range of the PIRs completely cover the area to be protected;
- identify the most vulnerable points, usually at the back of the premises, and ensure these are fully protected;
- don't waste money on a multi-zone system if it is not needed (but use more than one if the protected areas are well separated).

Second, it shows that 'experts' and installers cannot be wholly relied on. While many can and do give good advice and can plan a good reliable system commensurate with the risk factor, there are others who do not. Anyone with some knowledge of the equipment and a little electrical experience can set up as a security systems installer. But they may have little understanding of the overall security picture, its priorities and limitations of the various devices. Some will push for acceptance of the most expensive options irrespective of their suitability, as in certain other fields. Membership of trade associations so ostentatiously emblazoned on invoices and letterheads offers no guarantee to the customer at all; they are there to protect the the trader, not his clients.

The only way to avoid an experience such as the above is to carefully vet any proposed system with the aid of this book. Don't be afraid to question the 'experts'; after reading this book you could well know as much if not more than they do!

A passing comment is worth making on the strange behaviour of committees, perverse decisions being a well-known phenomena. Like individuals in a mob, and drivers when at the wheel of a car, sensible and reasonable people often seem to change personalities when working on a committee, and make decisions they would never make for themselves or if their own money was at stake. Sometimes a dominant member will sway the others, or a tentative comment by one is taken by the others as a concrete proposal and seized on. Whatever the precise mechanism, it is a fertile field for exploration by the psychologists.

12 Preventing shop-lifting

The major part of this book has been devoted to the prevention of burglary and with good reason, because in a typical year there are half a million break-ins with losses of over £70 million in non-residential premises.

But no less worrying is the prevalence of shop-lifting. Over a quarter of a million offences are recorded each year with a loss of some £10 million, but this is believed to be merely the tip of the iceberg, as by far the majority goes undetected. Surveys involving a number of retail stores having tight stock accounting systems suggest that up to one in every fifty customers steals some article and that the detection rate is as low as from one in a hundred to one in a thousand. This could put losses at a staggering £1 billion or more, far greater than that lost from burglary.

Surveys have also shown that many retailers are unaware of the extent of their losses through shop-lifting. Of total losses from all causes in one year, those surveyed could identify 9 per cent as due to shop theft, 2 per cent to employee theft and 89 per cent to causes unknown. It is clearly necessary to more accurately identify the causes so that effort can be directed where it is most needed.

A further point which emerges from these surveys, is that the popular conception of a shop-lifter as being a youth or a middle-aged housewife is not strictly accurate. While these are included, the ages spanned from 11 to 59 and no doubt there are many who practice shop theft who are outside those age limits. Much depends on the type of goods and their attraction to the particular age-group. Among shop-lifters at supermarkets, the middle-aged housewife may well figure prominently, but she is unlikely to prove a risk in the pop music department of the local record shop.

Prosecution

When a shop-lifter is apprehended, a decision is made by the store management as to whether the police will be called or the person given a warning. If the police are involved the result may be an informal caution by an officer on the spot, a formal caution at the police station, or a prosecution. If there is no prosecution, the store may bring a private prosecution, but if prosecuted, the result will be either guilty or not guilty. If guilty, the offender may receive a custodial sentence, a fine, a probation order or a conditional discharge.

From this sequence it can be seen that even if caught, the chances of an eventual deterrent punishment, i.e. a fine or custodial sentence, are slim compared to the various other options along the line. Added to this are the various circumstances considered by courts in mitigation, which are often presented by wily defendants with practised plausibility. Relevant to this is the criterion for prosecution given by the Attorney General which is:

> *'It is no longer satisfactory merely to have credible evidence to support a charge. The test now is whether a conviction is more likely than an acquittal before an impartial jury properly directed in accordance with law.'*

Cautioning by police as an alternative to prosecution has increased over the years, having doubled to an average of 42 per cent in one recent year, although variation between police forces varied from 16 per cent to 61 per cent. Apart from the odds against obtaining convictions with deterrent sentences, the Home Office treatment of offences for which a caution has been given as 'cleared up', offers police a strong motivation to caution.

Before the police are even called, store management have to weigh up the time and cost of pursuing the case against the value of the article stolen. In many cases, the theft of low-cost items is seen as unavoidable, and offenders if caught, are let off with a warning.

There are two unfortunate effects of all this. One is that the public and potential offenders are led to think that the penalties for shop-lifting are light and that it is easy to get away with. Many even consider that shops are fair game and that shop-lifting is not even criminal.

The other effect, is that those who are cautioned by shop managers may be habitual thieves who have been cautioned many times before at other retail establishments, and who therefore should not have escaped prosecution.

The only way this can be avoided is for retailers in the same area to cooperate by passing on information to each other concerning individuals they have cautioned. The only way this could work would be for names to be passed to a locally formed organization funded by the dealers. This would have to be done verbally by telephone: no list can be legally passed due to the over-protectiveness of the Data Protection Act. Each dealer could then make his own list of names, but other details, i.e. address, age, and ethnic origin should be omitted. The list should not be circulated to staff, but kept privately by whoever is responsible for security, for checking possible suspects.

Cost-effectiveness of security

While the reduction of shop thefts as a result of increased security is in itself highly desirable and likely to have a healthy effect on profits, it should, like any other business project, be balanced against the cost. The security could cost more than the losses!

Costs are not confined to the installation of whatever system is chosen, but must include maintenance, monitoring, staff training in its use, extra staff time taken in its operation, and possible loss of sales due to resulting delays in attending to customers. Display space and its effectiveness could also be adversely affected.

For example, electronic tagging is employed in many department stores. One chain of stores, after installing the system, recorded an average drop in shop theft losses of just under 50 per cent, although it ranged from 75 per cent to only 5 per cent in individual stores. The calculated cost of installation plus staff training was £68,000 for one medium-sized store. This did not include subsequent cost of staff time in tagging every item and removing them at the point of sale. The cost effectiveness could thus be questioned, especially for those stores achieving only 5 per cent improvement.

Closed-circuit television (CCTV) installations in another survey achieved a 25 per cent average loss improvement ranging from 75 per cent to just 1 per cent. The costs were £35,000 for a medium-sized store, but few subsequent costs. The installation costs seem rather high and could probably have been reduced and cost effectiveness improved if the system had been concentrated in the most vulnerable areas.

In a further study involving the use of CCTV in a single store having a high incidence of theft, it was found that not one of those caught was seen on TV, all were apprehended as a result of the vigilance of store detectives. However, the presence of the cameras undoubtedly deterred many and so played a part in reducing losses.

Another example of the hidden costs of security is what is known as the 'masterbag' display in contrast to the 'live' display system. With the former, goods are removed from their containers and stored separately, and the empty containers are displayed. Customers can examine the containers and take them to the pay desk if they wish to purchase, whereupon the containers are reunited with the merchandise. Live display has the complete goods on show.

This would seem to be the perfect answer for goods that can be easily parted and re-united with their packaging. It is commonly used in record and cassette shops where there is no need for the customer to examine the actual record or cassette, the sleeve or case gives all the information required.

It is less satisfactory for other products as quite naturally the customer wants to see what he is buying. Traditional counter service where an assistant produces the goods for the customer to examine before purchase, affords this. Although the advice and help of an assistant is sacrificed with self-service, the customer is compensated by the opportunity to choose and examine the merchandise at some length without the pressure to purchase. This advantage disappears with the masterbag display, and sales are almost certainly lost if the customer has only empty containers to examine.

Another factor is that storage space must be found for the unpackaged goods, and space costs money. Furthermore, the assistant must locate the goods and reunite them with their packets. This is time consuming and either means long queues at the checkouts or extra assistants for a given volume of sales. In fact it is almost back to the traditional counter service but without the service. So, what may seem a good idea could prove more expensive than the losses suffered by theft from live displays.

Crime analysis, the key

Blanket coverage of the whole establishment with high security methods or systems can be expensive and wasteful. While pilfering can occur in most areas and departments, it has been found that certain parts can be considered blackspots where a large proportion of the theft occurs. Concentrating the security in those areas will therefore be the most cost-effective.

This requires identifying the blackspots and analysing the pattern of thefts that are occurring. One method of doing this is by means of tighter stock accounting methods, the careful recording of stock movement between stores and shop, and identification of items sold. The latter presents the biggest problem. The sale of larger goods can be monitored by detachable labels, sales slips and receipts. Checkout tills that are operated by the bar code readily enable computerized checks to be made, but otherwise keeping track of small items is not so easy. However, some means must be devised if accurate analysis is to be made. It need be only temporary, for the purpose of the study, and later to check on the effectiveness of the counter measures.

Another method of analysis is by consulting past store records of those apprehended for theft. The reports of store detectives, if employed, are invaluable here. Points to be considered are:

1 The class of merchandise stolen.
2 The method of theft: by concealment, price label swapping, inclusion with goods actually bought.
3 If by concealment, where on the person, and if known, where in the store the items were secreted (those apprehended should be questioned on this).

4 Day and time of theft.
5 Class of offender: juvenile, teen-age, young adult, middle-age, elderly.
6 How detected: suspicious behaviour, observed concealing goods, technical detection (CCTV, electronic tagging), or other.
7 Factors that may have contributed: poor staff surveillance, secluded areas where goods could be concealed, stock layout (small high-unit cost goods in poor observation position or near exit).

All this information should be extracted from the records and presented in such a manner that the principal risk areas can be clearly identified.

It follows that detailed records should be kept of every apprehension. If such records are few, inasmuch as one in every fifty to a hundred customers steal, then a large number of thefts must have gone undetected and a security review is indeed urgently required!

Counter-measures

Staff awareness

One of the most effective measures, and also one of the least costly, is a staff trained to be aware of the problem. All too often situations are observed in retail shops and stores that seem positively to invite theft.

Training should include awareness of the old diversion trick and how to counter it. With this, two thieves, often school-children, work together; one creates a diversion while the other purloins the goods. Any such diversion should therefore be handled by one member of the staff, while all others are on 'red alert' and keep a more than usually watchful eye on other customers. Really, diversions can serve as a warning rather than a distraction to trained staff.

Instruction should also be given in the behaviour that has been observed in, and which can often identify, intended shop-lifters. The most common of all is looking about in all directions and at the nearest shop staff, usually in a furtive manner. An unobserved moment is being sought in which to steal. Genuine customers are more concerned with looking at the goods and the price labels.

Most customers examine an article for a few moments to ensure it is what they want before proceeding to the cash desk. Anyone picking up an article quickly, or taking an inordinately long time, though possibly genuine, could be intending theft, and so should be kept under observation.

Non-professional opportunist or impulse shop-lifters frequently display symptoms of nervousness. Among some that have been observed are sweating profusely, dropping articles, and fiddling with pockets, bags or clothing.

Persons wearing oversized top coats especially if unseasonable, carrying large open bags, or plastic bags with the firm's name, are all suspects and should be kept under surveillance. The latter is another old trick in which the bag having been obtained previously, is brought into the store to receive stolen goods. The idea is that anyone walking out of the premises with articles in a store's bag will appear to have purchased them and be above suspicion.

Staff should be well trained to recognize such tricks and signs and to be always vigilant. An incentive scheme such as operated by credit card companies to give cash rewards for each detection is well worth considering.

Store detectives

Store detectives have proved an effective means of detecting theft. While trained sales staff can reduce theft by vigilance, they have two disadvantages. Firstly, their primary task is to sell, so their time is mainly devoted to that and security must take second place. The second is that they are easily identified, so intending thieves can watch and wait until they are occupied.

Store detectives are primarily concerned with security and devote their whole attention to it. Furthermore, being dressed like ordinary customers they are more likely to catch thieves off-guard.

Cost effectiveness though must be considered. The time taken to deal with one offender, from the time of apprehension through reporting to management and interviewing, to accompanying to the police station and making a statement, can be some 2½ hours. This means no more than three arrests in a working day, even if a new offender was accosted immediately on returning to duty from the police station. The cost of employing a store detective must therefore be set against the value of the stolen items likely to be recovered.

On the positive side is the deterrent effect of known prosecutions by the store as a result of the store detectives. Also notices can be displayed that store detectives are employed, although these could be displayed whether they were or not.

Large firms should certainly employ store detectives. They can concentrate mainly, if not entirely, on those areas where the analysis reveals the greatest losses are occurring. With smaller stores the cost-effectiveness is likely to be less. Consideration could be given to employing detectives part time, if available, at busy times such as at weekends. With very small establishments, the floor area is small enough for sales staff, if vigilant, to keep under surveillance.

Bag parks

Customers are persuaded to leave bags at the entrance, for which they receive a ticket similar to a cloakroom ticket. This is done at some libraries and museums with success, as few people visiting these places take bags, and those that do may welcome being relieved of carrying them during their visit.

It has been tried at some stores but the situation is very different. A shopping bag likely will contain a purse that the owner will need and otherwise would have to carry in her hand. There may be other valuable purchases that could easily be taken from an open bag in the park. With a large number of bags arriving and leaving, mistakes can easily occur, and at busy times there may be queues to deposit and collect. Such an arrangement is very likely to inhibit genuine customers from visiting the store and is not recommended.

Electronic

CCTV

This is the first electronic countermeasure that usually comes to mind. While CCTV can provide useful surveillance to detect intrusion into premises, it is less useful in detecting shop-lifting. The minor quick action needed to conceal a small item on the person or drop it in a bag, is not easily seen on a small monitor especially if the offender is some distance from the camera. He could also walk out of range with the goods in his hand as if making for the cash desk, then conceal them in some blind nook or corner. Furthermore, full-time monitoring rather than a casual glance is required in a busy store.

While CCTV is poor for *detection,* it rates much higher for *prevention.* The very sight of cameras is a big deterrent. This being so, many stores have dummy cameras installed. This is quite a valid stratagem, *providing they look realistic.* Some dummies look just what they are, especially those having a large number of 'lenses' protruding from a small hemi-spherical base. Many youths who are among the major offenders, are familiar with the appearance of video cameras and are not easily taken in by poor imitations. A dummy should look real and have both a mains and video cable coming from it. These can disappear into a hole in the ceiling and go no further, but they add realism.

A cost effective system could have a few real cameras at especially vulnerable points with at least one monitor in public view, with a number of dummy cameras scattered around at other points. Some of these could be located at those blind corners, stairways and so on.

Electronic tagging systems

These systems have been mentioned before. A special tag is fitted to all articles to be protected and can only be removed with a special tool at the checkout. Any tag that passes detectors situated at the store entrance triggers an alarm. This is probably one of the least cost-effective measures.

It requires much staff time in tagging, then in removing, and is expensive to install. For small-volume expensive items it is probably a good idea, but for large-volume inexpensive articles, the operating costs would most likely outweigh reduced losses.

Another factor is that with a number of people passing in and out of the entrances during busy periods when theft is most likely, there is no way the offender can be identified. All cannot be detained, and if the thief keeps cool and the item is well concealed, he can probably just walk away. If an alarm is sounded he would probably be well away anyway by the time a store detective reached the entrance. Really, the system requires a security man to be permanently stationed at the entrance thereby adding to the operating cost.

Display alarms

This type of alarm can take various forms and can be very cost effective. It may be thought that large items such as TVs, microwaves, video recorders, lawn-mowers and the like, are too large to steal from a showroom. Dealers who stock such items will soon tell you otherwise. A different technique is required, instead of the furtive look around and a quick slip into a bag or pocket, the thief is bold, walking coolly out in full view with the item. Sometimes a warehouse coat gives the impression of being on the staff especially if of the same colour as those issued by the firm.

Fortunately, protection is easier than for smaller goods as it is not usual for customers to take large items from the display to the cash desk. The displayed items can therefore be immobilized by a display alarm. One form of display alarm is the linking wire (Figure 60). This consists of a single wire that is passed through handles, rings or other enclosed projections on the articles to form a complete loop back to the control box of an alarm system. The terminals can be fitted at some convenient point on the display, and the wire loop connected or plugged in. It can be broken at intervals so that it is made up of sections terminated with small plugs and sockets such as the 2.5 or 3.5 mm audio jack. This enables a single item to be removed from the display by an assistant without unthreading the whole loop.

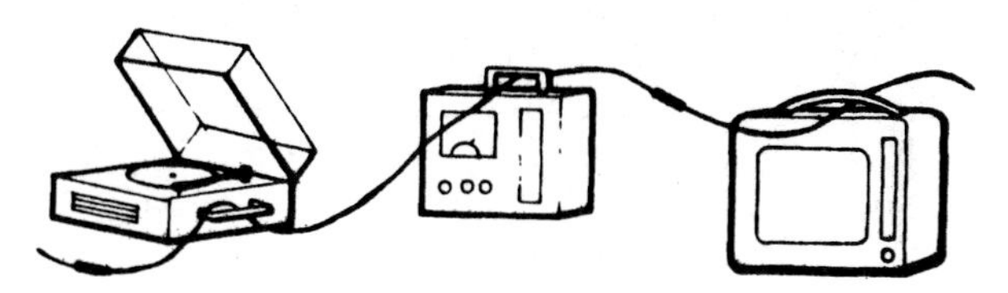

Figure 60 Wire loop threaded through handles or other projection of displayed merchandise. Small plugs and sockets (2.5 or 3.5 mm audio jacks) enable items to be removed by the staff.

A bypass switch can be fitted at some suitable point only accessible to staff, or it can be a key-switch. The loop can then be broken to remove an item without sounding the alarm. If the loop is a large one covering several displays, sections can have their own individual bypass switches (Figure 61), so that a section can be opened without inactivating the whole loop. The control unit can be a simple affair which latches on when the loop is broken and sounds one or more buzzers at appropriate locations.

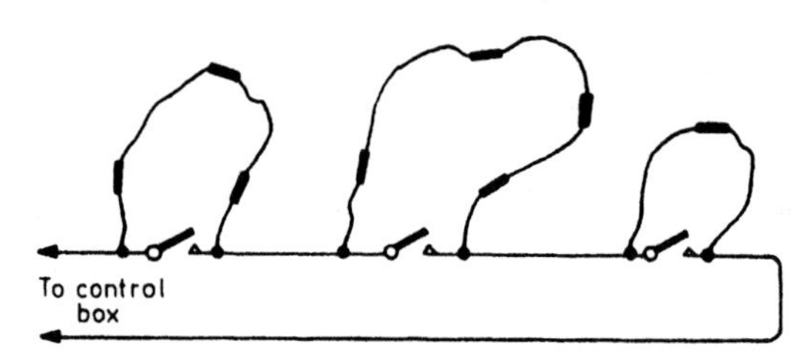

Figure 61 Several displays can be served by the same loop, but each should have its own bypass switch so that all are not deactivated when it is required to break any one.

A rather ingenious alternative consists of wiring a number of mains sockets in series and mounting them along the back of the display stand or window (Figure 62). These thereby form a loop which is connected to a similar control unit as before.

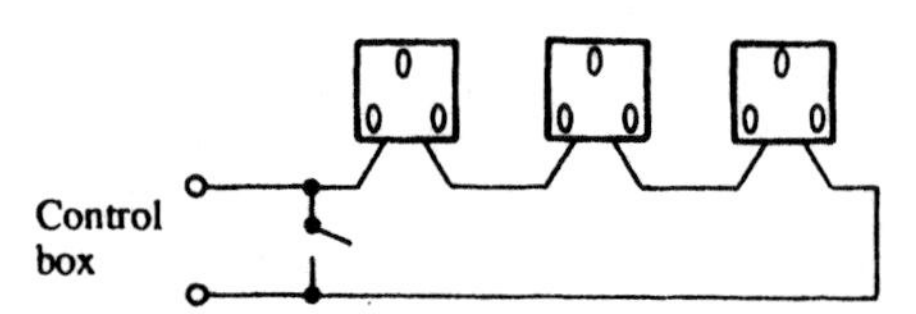

Figure 62 Mains sockets connected in a detection loop for electrical goods display. Appliances must be switched on. A concealed bypass switch is fitted to permit removal by sales staff.

Appliances are fitted with mains plugs wired in the conventional manner, plugged in to the loop and left switched on. The internal transformer or motor offers a low impedance to the small d.c. current passing along the loop and so completes the circuit. If one is unplugged or the mains lead cut, the circuit is broken and the alarm is triggered. A bypass switch must be fitted for each bank of sockets to allow an appliance to be unplugged by a member of staff.

The advantage of the mains plug system is that threading, connecting and disconnecting of the linking wire system is avoided, and also the possible inadvertent disconnection caused by a customer examining an article. One small disadvantages is that all sockets in any bank must be filled to complete that section, but if banks are kept small, say to three sockets, there is little inconvenience.

Pressure mats

These can be used to protect larger, heavier goods displayed where they could be stolen, such as near the showroom door. They can be deployed under carpeting in the usual manner, but instead of keeping objects clear, the displayed articles are placed on them. This means that their contacts are normally closed, so unlike the normal mode of connection they are wired in a loop. If the protected article is removed the mat goes open-circuit and the loop is broken.

The pressure required to operate a mat is around 3 lb per in^2 (211 g per cm^2), so the displayed goods must exert at least this pressure to close the circuit. If the base area divided by the weight of the article is less than this the pressure will be insufficient. This can be overcome by using a small tripod as a display stand and standing one or more of the legs on the mat. Owing to the small area of the foot the pressure will be much greater.

Thus a tripod having feet with an area of 1 in^2 each has a total floor contact area of 3 in^2, therefore an object of 10 lb (4.5 kg) exerts a pressure of 3.3 lb (1.5 kg) on each foot which should be sufficient to make contact (Figure 63).

Figure 63 A pressure mat wired in a loop (it is normally closed when article is in place) can be used to protect larger items. If placed under the leg of a display stand, the weight is concentrated in a smaller area and pressure is therefore greater. Otherwise weight may be insufficient to operate it.

Microswitched stands

A possibility for light objects is the use of microswitched stands (Figure 64). The top of a box-like structure is hinged on one side and supported on the other by a microswitch. Its weight is insufficient to depress the switch, but any article placed on it adds its weight and the switch is held down. Should the article be removed, the switch is released and an alarm triggered.

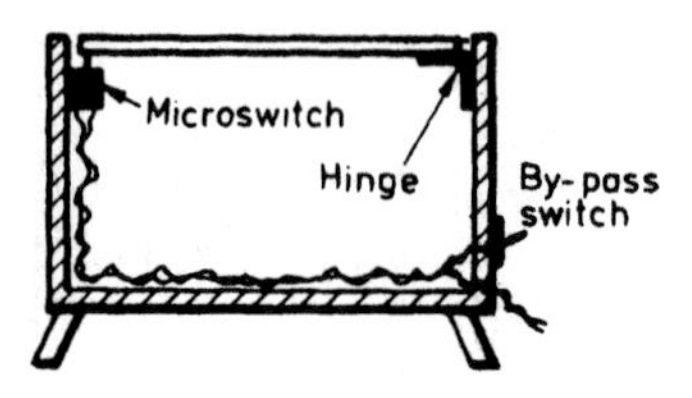

Figure 64 A display stand can be made consisting of a box with a hinged lid. A microswitch supports the lid on its free side, and is depressed when displayed article is in place. When the article is removed, the switch is released. A bypass switch at the rear or underneath enables the staff to remove goods when required.

As most microswitches have single-pole double-throw (SPDT) change-over contacts, the switch can be wired either to open a loop, or close a circuit. It could thus be wired with a battery to actuate its own buzzer, the buzzer and battery being fitted inside. The stand would thus be self-contained and usable anywhere. A bypass or on/off switch can be fitted unobtrusively to the back or underneath.

Exit barriers

Fire doors, through which offenders could easily depart with their loot, require exit barriers. Yet they must not hinder an emergency exit. A solution is the exit alarm shown in Figure 65.

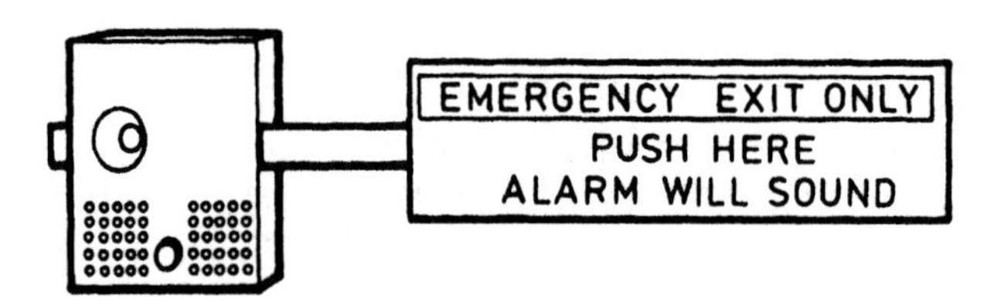

Figure 65 Exit alarm to prevent unauthorized use of exit doors without restricting emergency use.

Normally, fire doors are secured by the standard panic bar comprising a horizontal bar across the door which releases bolts at the top and bottom when pushed, so allowing the door to be quickly opened from inside but securing it against intrusion from the outside. The exit alarm consist of a security bolt fitted to a steel box which is mounted on the door. It is released by the push-bar but this also sounds an alarm. A notice to that effect appears in large lettering on the bar. Authorized staff can bypass the alarm by use of a key. The alarm and battery is contained in the box so the unit is self-contained.

Another solution to the fire exit alarm is to wire it to the 24 hour loop of an intruder alarm. A notice on the door should warn that opening it will sound an alarm.

Mirrors

A popular device often seen, is the convex mirror placed at some strategic position. Its feature is that it covers a wide area, and offers observation from a range of locations. Thus staff can see a large section of storespace from various positions. It also is meant to be a deterrent, warning potential thieves that they are likely to be observed.

It can, however, be somewhat of a two-edged sword. First, because the coverage is over a wide angle, the image of any particular person is small, just as it is with the CCTV monitor. So any suspicious actions could easily be missed. Second, it also works the other way, the offender can keep an eye on the proximity of sales staff and whether they are looking in his direction. So it helps the hunted as well as the hunters, and therefore is of doubtful worth.

A far better proposition are one-way mirrors. These can be installed on walls dividing the store from offices, stairways, and any adjacent rooms. Prominently displayed notices should warn that one-way mirrors are in use and detection of thieves is certain. The offender has no idea if there is anyone behind the mirror or whether he is being observed or not. Unlike the convex mirror it does not help him at all. As visibility through these mirrors is not restricted to a narrow angle, he cannot even move sideways out of range as he could with a CCTV camera. To add to his uncertainty, a number of ordinary mirrors can also be mounted in various positions. There is no way these can be distinguished so he is given the feeling that eyes are watching everywhere!

If the store is long and narrow, the walls, and therefore the mirrors, are not very far from any display, so surveillance or suspected surveillance is all around. There is no safe place anywhere to secrete stolen goods.

With a large square floor area the walls are more distant, and the thief may feel safer toward the centre. There are two possible answers to this situation, a walled structure with mirrors could be raised in the middle of the floor to house an observation point, office or storeplace. This would reduce the distance between any pair of parallel walls to less than half.

The other is to arrange the displays so that the smaller high unit-price articles, or those which the analysis revealed to be popular with thieves, are in full view of the mirrors, while the larger or less desirable items are further from them.

This really could be classed as a separate counter-measure. Whatever means of surveillance is used, the vulnerable stock should be located where it is most effective, and only those items unlikely to attract the thief should be in the furthest reaches where observation is difficult.

Cost-effectiveness of one-way mirrors is high. Although not cheap to install, they compare favourably with CCTV, electronic tagging and many other measures. Once fitted though, there are no maintenance costs other than occasional cleaning, and no staff time absorbed in using them, unless someone is stationed in that central lookout.

It would perhaps be rash to point to any one counter-measure as the best and most effective of all; each has both advantages and drawbacks, and a combination of several is likely to prove most practical. If one could be singled out after weighing up all the features of cost-effectiveness, minimal staff-time absorption, convenience, practicability, no maintenance, offering both high detection and high deterrence, the one-way mirror system would almost certainly be the winner. It is therefore strongly recommended.

Finally, let us end this chapter on a more positive note. If it depresses you that at least one in every hundred people that enter your store come to steal, remember that ninety-nine are honest customers who have come in to spend money!

13 Violence against staff

It is regrettable that violence against employees in shops, offices and other premises open to the public is becoming an all too common fact of life. In the years 1993–4 12,055 physical attacks were recorded, 90,421 were threatened with violence, and a staggering 209,645 reported verbal abuse. In 1993, 1 in 5 retail premises were the scenes of violence, although most of these, as the figures show, were verbal. It is believed that many more went unreported.

According to the Health and Safety Executive, violence against employees is defined as 'Any incident in which an employee is abused, threatened or assaulted by a member of the public in circumstances arising out of the course of his or her employment.'

What type of establishments are least and most vulnerable? Furniture shops are the least, having no recorded incidents in the period surveyed. Books and stationers are very low, then jewellers, food retailers, Post offices, chemists, department stores, and petrol stations, in that order. At the highest, way above any other is the off-licence, perhaps understandably. Below this in the high category, are DIY and hardware stores, general retail, and music shops and about half-way up are footware shops.

There are many different reasons for attacks on staff. The largest, accounting for 49 per cent is the prevention of theft. Next is in the course of robbery at 19 per cent, then general troublemakers at 15 per cent. Persons under the influence of alcohol or drugs account for 9 per cent and finally angry customers with some grievance account for 8 per cent.

Theft prevention

Prevention is better than cure, and the points made in the previous chapter should go a long way to deter the would-be thief from even trying. However, if detected in the act, the situation can then become hazardous for any sales staff who have to apprehend the culprit. Violence could result.

A calm approach is the best, avoiding the use of emotive words such as 'theft', 'stealing' and the like. In a large store security staff are employed to deal with the situation, and they should be summoned straight away. In the absence of security staff, a policy should be established by the management as to how detected shoplifters should be handled. Pursuing a fleeing thief into the street is not usually a good idea, as if caught he is likely to become violent, and rarely will help be forthcoming from the public, in fact they frequently go to the aid of the offender.

When accosting the person, the staff member should be observant, making a mental note of anything and everything about him that could be used to describe and later identify him. A general observation is not really sufficient, get into the habit of analysing a person's features so as to describe them accurately and build a photo-fit picture if necessary. Look at the hair, its colour, style, and quantity. Next the forehead, is it high or low, is it marked? Next the eyes, are they widely set or narrow, and what colour? Is the nose long, short, turned up, bent, wide or narrow? Is the upper lip full or narrow? Is the chin rounded or pointed? Are there any scars or marks on the cheeks? Is the shape of the face round, oval, or heart-shaped? Any face jewelry? Are the ears flat or protruding, normal or misshapen?

Getting a detailed 'fix' of the face is far more important than describing the clothes, which most victims do. These can be changed, and usually are nondescript, there being thousands virtually the same. Height is another thing that can narrow identification. The best way is to be aware of your own height, then estimate how many inches the offender's eyes are higher or lower than yours.

Making a swift memory picture in this way needs practice, and it can easily be tried out on ordinary customers – without their knowing what you are doing of course.

Robbery

The difference between robbery and theft is that there is no violence at the actual time of a theft although it may occur later if the offender is apprehended. Robbery occurs when violence or the threat of it is part of the act. Thieves do not usually intend violence, but hope to get away unobserved. Robbers use violence or the threat of it as their stock in trade. Thus a robber is usually more dangerous than a thief.

The best advice is not to resist, but hand over what is demanded. Do not 'have a go'. A life is more precious than any amount of money. A panic button should be placed within reach of a cashier in a position so that the robber cannot see it being operated. At foot level is an obvious place, operated without too much foot movement. It could be set to ignore a single press but operate on two. This would avoid false alarms from it being accidentally kicked or trodden on.

If the alarm is raised in the office or security HQ, and is not heard in the shop, this will add to the safety of staff. An alarm suddenly sounding could panic the robber who could then attack the operator in revenge, or worse still use a firearm, before getting out. Better for the office to call the police who may be able to catch the villains outside if they are quick enough. To facilitate this the cashier could cause a delay by dropping things, being slow, or some other ploy without making the delaying tactic obvious.

As with the thief, identification is more important than attempted capture. Likely the robber will be wearing a mask that will make facial features unrecognizable, but there are other things. Is the hair visible? What about the ears, earrings? Height can be determined, and colour from exposed body parts such as hands, or wrists if gloves are being worn. Look for rings, watches or bracelets. What is his build, thin- or thick-set? Did he have any accent (although this could be disguised)? Are the shoes visible? These are not usually changed as often as outer garments and so could provide a clue if unusual.

Where video cameras are installed these will do the job to some extent, but the quality is notoriously poor and a good eye-witness description will do much to augment the camera shots. It is understandable that fear may paralyse one's power of observation, but if it can be trained in advance, it will come naturally.

Robbery may not always be for money; staff could be threatened into handing over goods and so may not be near a panic button. The action to take in such a case depends on the circumstances. Is the robber armed? Does he seem to pose a real threat? Whatever the situation it is best to be on the cautions side; he may have a concealed weapon. But sometimes a blunt refusal will leave the offender confused and he will just walk out. Be careful though.

Troublemakers

These comprise the next frequent cause of staff violence. Youths are often responsible. If several enter together it would be wise for more than one staff member to be present nearby to handle any trouble. Watch out though for the odd one who may detach himself from the main group and browse at another part of the store. This is a frequently-used trick to distract attention from the shoplifter who makes a good haul while staff are keeping an eye on the rowdy act elsewhere.

Many stores make it a rule not to admit more than one child or youth at a time, and a notice is displayed at the door to that effect. Should several then try to gain entrance, the notice can be pointed out in a non-confrontational manner. Something along these lines: 'I'm sorry we only allow one young person in at a time. You are probably quite honest, but we do get others who are not, so we have to make the rule and stick to it for everyone.' Notice that no imputation of wrong motive is made in their case; the rule is due to the dishonesty of others. Hardly provocative. The suggested statement is just basic, it could be padded out as felt necessary.

As is often the case in other areas, the innocent may suffer because of the guilty. The group may be genuine, several friends going along with one to make a purchase, and who may be upset at being considered potentially dishonest and not allowed in. However, a smiling, friendly, and reasonable manner on the part of the staff member should dispel resentment.

Others may be bent on trouble for its own sake. Sports fans at the venue of a big match are a notable example, especially if their team loses. As little business is likely to be done on such occasions, probably the wisest course is to put up the shutters and close the shop for the afternoon.

Drunks and drug abusers

Some 9 per cent of physical attacks are from people under the influence of alcohol or drugs, so although a minor proportion, it is still significant. Sometimes reasoning in a friendly manner and escorting them to the door is all that is required. Others who are showing belligerence should be left alone and a call made for the police to deal with them. Several staff members should remain nearby until help arrives in case the person becomes really aggressive.

Angry customers

Although angry customers constitute the smallest proportion at 8 per cent, they should be handled quite differently from the other groups who are out to steal or otherwise make trouble. You will be glad if they never set foot in your premises again. Angry customers though *are* customers, so the object is to pacify, help, reconcile, and hope they will continue to patronize your business. Apart from loss of their custom, they will likely broadcast the incident to friends and neighbours, so an unfavourable outcome could mean the loss of much more business. Bending over backwards to satisfy will engender goodwill, and this too may have a knock-on effect.

Most angry customers have, or believe they have, a legitimate complaint about goods or services supplied. An obstructive or unhelpful attitude will definitely put you in the wrong in their eyes and may lead to acts of violence, or at least threats.

Whatever the complaint, staff should be instructed to take it seriously, listen carefully without objecting or interrupting, ask helpful questions as to the whys and wherefores of the alleged failing and apologize for the inconvenience caused, without at that stage committing the firm to any liability.

Unless the matter is a trivial one which can be dealt with there and then, it is wise to call in the manager or supervisor. Thus the person who is in a position to do something about it is involved, and the customer will feel that his complaint is indeed being taken seriously. Actually few people like complaining and most would rather suffer a loss, so those that do pluck up courage to go back and make a complaint should be put at their ease and not made to feel small.

If the complaint is genuine, then profuse apologies should be made, the matter put right immediately if possible, and perhaps some additional compensation made for the customer's trouble.

Some customers may become angry at their first visit because of the treatment they receive. Someone who has a limited time left on his parking meter, and who is kept waiting while one assistant is serving and one or more others are carrying out some duties that could wait, such as stocking shelves, arranging displays, or, as is often the case, just talking to each other, is likely to erupt! Unnecessary waiting is the cause of much irritation. Another cause is the attitude of the assistant; a blank-faced, terse 'Yes?' is not the greeting likely to promote good feeling. Minor irritations perhaps, but they could build up, especially if some other annoyance has happened before entering the shop.

All staff should be instructed that attention to the customer is first priority, no-one should be subjected to needless waiting. Other tasks can be left in order to serve, unless of unusual urgency. Furthermore all customers should be greeted with a warm smile and pleasant manner, but not a silly grin which could be misinterpreted.

There are of course the angry customers that are trying it on; they are not genuine and they know it. They are out to defraud by returning goods damaged by themselves, or even returning goods for a refund that have been stolen. With borderline cases where the fault may have been present when sold, the customer could be given the benefit of the doubt and some reparation made. However, do not go too far down this road, otherwise your business will soon become known among the criminal fraternity as a soft touch, and you will be getting many more such cases.

If it is evident as a result of your security measures that a claim is fraudulent, again the staff member should call on a supervisor. Two will thus be present should violence erupt, but in fact it will be less likely. The offender should be told firmly but non-aggressively, that your security measures have identified an attempted fraud, and any further attempt will involve calling the police. Of course it may be your policy to call the police on the first attempt, in which case the offender will have to be delayed on some pretext. If a video camera is in operation, the appropriate part of the tape could be printed or otherwise preserved for future identification.

Credit card fraud

Staff will be well informed by the credit card companies of the signs to look for when a credit card is being misused or has been stolen, so these do not need to be repeated here. However, any suspicion which may result in the need to retain the card, could certainly lead to violence. Once more, any such suspicion should involve calling a supervisor to make the decision to retain. Where there are two or more staff members present, the possibility of violence is always lessened.

Threats of violence

Threats have been classed with actual violence, but some may feel that they are of minor importance. To some maybe, but particularly to female staff, they could be very disturbing and upsetting, especially if made in a very aggressive manner. They thus should be considered as a violent act and logged or reported as such if records are kept, which they should be of all violence against staff.

Security guards

Most large stores have uniformed security guards, and these indeed are a deterrent both for shoplifting and acts of violence against staff. Smaller firms may not be able to afford one, but some small firms with nearby premises have combined to hire a guard who patrols all the premises, spending a certain specified time in each, and who is on call via a pager should trouble crop up in any one.

A number of security firms can supply guards, but not all such firms hire desirable staff, some even have criminal records! There is a British Standard for manned guards – BS 7499 – so inquire whether the security firm meets that standard and have thoroughly vetted their staff. Are they insured against any losses due to negligence or dishonesty on the part of their staff? A word with the local crime prevention officer at the nearest police station may be worthwhile to direct you to a reliable firm, although it must be said that some police forces are not greatly in favour of *any* outside security guards. Professional jealousy perhaps.

Business co-operation

Apart from sharing a security guard, there are other ways that traders in any locality can help each other in security matters. Persistent trouble makers get known, and if any appear in the vicinity, a phone call around to neighbouring businesses warning them would be helpful to all. Any new stunts pulled by ever devious and inventive criminal operators should also be communicated quickly to fellow traders.

14 Employee theft

Burglars call once, and if the security is good, get away with nothing. Shoplifters may come several times, but if security is well planned, are likely to be deterred or detected. Dishonest employees come every day, know the ropes, are trusted, and can bleed a firm almost dry. More than one business has gone into bankruptcy because of sustained employee theft.

Of all types of loss, employee theft is not only potentially the most damaging, but perhaps the most saddening because the offenders are persons who are known and trusted, who are given access to goods and property that is guarded from outsiders, and whose own prosperity depends on that of the firm.

Statistics reveal that some 20,000 employees are prosecuted each year with an average for each case of over £900, a total of over £18 million, double that from shop-lifting. Surveys show that while some 60 per cent are referred to the police, 40 per cent are warned or disciplined by management, so this could mean a detected loss of £30 million. Undetected losses are likely to be far greater, some estimates put it as high as over £1 billion. A Gallup poll revealed that only 8 per cent regarded theft at work a crime worth reporting to a superior, and a majority did not consider it a crime at all – an astonishing revelation.

Petty pilfering, too, can mount up into sizeable amounts, yet in the prevailing ethos is considered not dishonest at all but as one of the 'perks' of the job. Persons who would never dream of committing burglary, mugging, or stealing from workmates, will purloin articles from work without compunction.

An anecdote illustrates the point. An irate father berated a teacher at his son's school because other children were continually stealing his pens. 'It's the principle of the thing I don't like,' he declared, 'it's not the cost of the pens, for I can get plenty more at work.'

Many managements turn a blind eye to petty pilfering thinking it would sour staff relations if prevented. Some allow it because otherwise a finger may be pointed at their own fiddles. Whatever the reason, condoning petty pilfering is unhealthy, it encourages dishonesty and promotes an atmosphere in which more serious theft becomes that much easier. It is far better to stipulate that certain items are available free or at reduced cost in reasonable quantities providing a supervisor is asked first. This keeps things under control and as a bonus, generates a liberal image.

It is sometimes assumed that theft is committed almost exclusively by sales staff or others on the lower rungs of the ladder. Management, may be regarded as above suspicion, but surveys have shown that theft is not confined to one class of employee.

In one survey, the proportions of offenders worked out at: managers, 8 per cent; supervisors, 11 per cent; sales staff, 55 per cent; and non sales, 26 per cent. While this 19 per cent of managers and supervisors is itself disturbing, the figures are even worse when it is considered that they are fewer in number than sales staff. So if there are five sales to one supervisor, and seven sales to one manager, the proportion of dishonesty among management and staff is equal. A larger number of sales staff tips the scales the other way so that there is actually more dishonesty in management.

Methods

As with shop-lifting, the first stage in prevention is knowing the extent and methods of implementation. Much can pass undetected or be attributed to customer theft. The analysis recommended in the previous chapter to trace stock movements should help to some extent, although it may not be possible to entirely distinguish between customer and employee theft. Disappearance of stock before reaching the showroom would certainly be attributable to employees. It may not even arrive at the stores due to collusion between staff and delivery drivers.

If we trace the various points at which theft can be committed, the non-arrival is obviously the first one. The next is stealing from the store. This in many establishments is the vulnerable point, because once stock is received it is placed in store to be drawn on as required with no further paperwork. But a record should be kept of stock movements out of the store so that any losses therefrom can be identified.

Next, stock can disappear in transit between stores and showroom. Here again, if suitable records are kept, goods booked out from the store should tally with those booked into the showroom. Theft from the showroom is perhaps the most likely, as there are more distractions and opportunities, and it can easily be disguised as customer theft.

Lastly, goods can disappear at the checkout by what has come to be known as 'sweethearting' for friends and relatives, when the assistant enters a smaller amount than the actual price, or does not enter some items at all.

Just taking money from the till would create a deficit and soon raise suspicion. But cash can be stolen by underringing purchases while charging the correct amount to the customer, or ringing the correct amount and giving wrong change. Later, the surplus cash is removed from the till so that it balances.

An inexperienced supervisor may think that while a till balance in deficit is cause for suspicion, a surplus must just be due to error. A surplus should be regarded with equal concern as a deficit. It could arise from the cashier losing count of overcharging and underchanging, or from having had no opportunity to remove the surplus.

Modern tills make cheating difficult because totals and the amount of change are displayed for the customer to see. However, many people fail to check their change, especially when the store is busy. Opportunities thus exist.

Other dishonest practices are the abuse of discount privileges, that is selling goods to friends acquired at discount for personal use; abuse of staff damaged or soiled stock discounts, whereby wanted items are deliberately damaged or marked; and fraudulent use of credit vouchers.

Removal

Having stolen the item, the next step is to remove it from the premises. If small, stolen articles can be concealed in bags, pockets or otherwise on the person in much the same way as does a shop-lifter. The employee has good opportunity for concealment on the person during visits to the wash-room. Too-frequent visits could thus be a warning sign, and may warrant putting the person under surveillance.

Concealment in personal lockers or depositing in some location from which later collection is intended, are further methods of getting the goods out. One often-used method is concealment among rubbish. When rubbish is put out overnight to be collected early next day, stolen items are placed in the refuse, and then picked up later in the evening.

Most stolen goods are carried out through the main or staff entrance when leaving, but larger items may be taken out through goods or other entrances. These may be taken unobserved to the entrance by the employee, and picked up by a waiting accomplice outside. This eliminates the problem of the employee having to smuggle the article out when leaving, without arousing suspicion.

Management

Administration staff have more opportunities for dishonesty than most others. For one thing supervision over them is limited. Also, as they are in control, they can adapt procedures to provide favourable situations. They usually prudently operate within realistic limits, so that losses do not attract undue attention and can be attributed to other factors. As with the lower-status staff, collusion may take place, or there may be just one person involved. There could also be collusion between management and suppliers' agents to divert stock, or authorize payments for goods not received.

Cash takings may shrink due to management handling, and unless the accounting system is tight, there are obvious opportunities for this as cash cannot be traced.

Company fraud

If the losses suffered from employee theft of goods seems high, those arising from computer fraud and fraudulent cash conversions are horrendous. They are estimated at some £1.5 billion a year, and some authorities consider even this to be an underestimate. Among other practices brought to light have been workers in manufacturing and service businesses running 'parallel businesses' using the firm's time, materials and equipment, and poaching the firm's clients.

Modifications to computer entries is both easy and widely practised as many who operate them have acquired quite advanced computer skills at school or as a hobby, and soon find ways and means of putting them to their advantage.

Information leakage to rivals or others who may use it to their advantage is another common practice, with employees often being all too ready to pass on information for gain.

Most of these company frauds have only been discovered by accident rather than as a result of surveillance or preventive action. Undoubtedly many more, perhaps the majority, remain undiscovered. Expert investigators have expressed amazement at the lack of preventive measures which so often enable large-scale frauds to pass undetected. The crimes, they have found, are more often than not committed by persons who have a 'predisposition to steal', a significant point that we will take up later.

The counter-measures which follow are intended as a guide for the prevention of all the foregoing classes of employee theft.

Counter-measures

For large companies, 'ghost employees' are a possible source of loss. Do not take the payroll figures for granted. There should be dual control of payroll, and a periodic check made by someone in higher management that the names on the book are of people that actually work there. Check also the work rates. In one firm the night shift applied for extra time when the clocks went back, which it would seem they were entitled to. However a check on the previous six-months revealed that the same team were on duty when the clocks went forward but they were paid the same as a normal shift. Needless to say they did not get the extra hour's pay!

Most firms accept returns for faulty or damaged goods. Check that the number of returns are at a reasonable level; if they seem to be too high, check it out, get payments for returns authorized by senior management. If the number of returns then falls it is obvious that false returns have previously been paid for.

Petty cash may not be so petty. Quite large sums can disappear over a period. Every item should be entered in a suitable book, even in a small firm, and the recorded expenditure tallied with the cash paid out. The book should be audited at regular intervals.

Fiddling the books has often been discovered when the cashier has been unexpectedly away. A dishonest employee will endeavour to keep the books out of sight of anyone else, at least for such a period that may lead to discovery. In these cases holidays tend to be taken in short breaks. If any employee responsible for cash, payments – or who has any other opportunity for fiddling – always takes their holidays in short stretches, be suspicious. Make sure they take a fortnight in one stretch, and make a thorough check in their absence. Be suspicious of outsiders, travellers, drivers, or others who call and enquire after them while they are away – they may be part of the fraud.

Tight stock control

Tight control of stock from the time it enters the stores, ensuring that it does actually enter the stores, at every step right to the point of sale, is an essential measure. In multiple-outlet firms the procedure should be designed by Head Office for uniformity, and so that no-one can think they are being singled out for special attention. Furthermore this removes the possibility of local management tailoring the system to suit their own purposes.

Cash control and audit

Occasional errors in till balances can and do occur, but frequent discrepancies whether surpluses or deficiencies in one particular till, or tills operated by one particular cashier, should be viewed with suspicion and surveillance instigated.

A means of recording the actual total for each till and identifying it in the total takings that are banked, should be instituted. In addition to the normal branch audit, unannounced spot-check audits should be conducted from time to time. All stock movements and cash accounts should be scrutinized.

Electronic surveillance

The same closed-circuit television (CCTV) cameras that deter shop-lifters can do the same for dishonest employees though to a lesser extent. Unlike the public, they will know which if any are dummies. One-way mirrors overlooking the sales area from the office or other administration rooms can also prove a strong deterrent.

A camera over the checkout area could deter sweethearting as then no cashier could be sure of being unobserved. Another at the loading entrance could prevent goods going out that way, and also would give an apparent check on irregular deliveries.

These surveillance measures may cause resentment in some employees who may feel that they are being spied on and are not trusted. It should therefore be made clear that they are part of the general security system to prevent theft and deter possible attack on staff members. They are thus there for the benefit of all. A parallel could be drawn with the video systems that have helped to identify bank robbers.

General surveillance

Store detectives should be reminded that it is also part of their job to detect employee theft. While their primary concern is the detection of shop-lifting, and not that of watching staff, they should be alert for any tell-tale signs pointing to staff dishonesty. Among these could be frequent visits to the wash-room, unnecessary visits to the stores or prolonged stays there, loitering near exit doors, nervousness, or any of those signs commonly observed in suspected shop-lifters. Supervisors should also be on the alert for these occurrences.

A watch could also be made for suspicious-looking parcels, bags, or objects concealed under coats of employees leaving the premises. Surveillance of the staff car-parking area may also be revealing. Stolen goods are likely to be brought out of concealment, and stowed away in the vehicle.

The degree of surveillance would to some extent be governed by the results of the crime analysis. If there seems no problem, surveillance can be light, though never completely relaxed; but if employee theft is apparent, it must be intensified. A lax attitude on the part of management only encourages offenders to continue and likely increase the amounts stolen; it can also encourage others to follow suit.

Personal identification

Where computers and information systems are at risk, measures are needed to ensure that only authorized staff have access to them. While they could themselves be dishonest, thorough screening of employees for such positions of trust as described later, should reduce the risks.

With many of the successful frauds investigated, there was little or no restriction of access to computer terminals. It follows that keeping out low-security-grade employees will greatly minimize the chance of fraud by this means.

Various means exist for limiting access: keys, passwords, and code numbers all can be used to good effect. However, with such large sums at stake, all of these must be considered of comparatively low security status. Keys can be stolen temporarily, duplicated and then replaced without the holder being aware of it. Passwords and codes can be discovered, if only by covertly watching them being entered by an authorized person.

One interesting method of verifying identity is the electronic signature analyser known as Securisign. Personal signatures contain many distinctive features. These are not just the geometric appearance of the signature, but the pressure exerted at different places, the rhythm, the angle the pen is held, and even the flourishes of the pen which do not actually appear on paper. While a skilled forger can make a good visual copy of a signature it is very difficult if not impossible to duplicate it in every respect.

With this device authorized staff make their signatures using an electronic pen, and the associated microprocessor extracts thirteen different features from each and stores them in its memory. When verification is needed all they have to do is to sign with the electronic pen, and the circuits compare the result with the stored original.

Because no one signs their name exactly the same each time, the device looks for the important constants in the signature and gives these greater weight than any more superficial variations. The recorded signatures in each Securisign unit are customized so that signature records from one unit would not function in another.

The device also provides protection against 'hackers', and enables the 'disclosure' provisions of the Data Protection legislation to be complied with and many other applications where the identity of someone not personally known and where contractual signatures must be verified. Further details can be supplied by the maker, A1 Security, Cambridge CB2 4EF.

Good staff relations

Theft is often committed by staff who feel they have a grudge against the firm or the management. Stealing, is a way of hitting back, or of 'compensating' for deprivation of what they consider to be their just dues. Of course there will always be some who will steal whenever they get the chance, whatever the relationship. However, employee theft is undoubtedly higher when the staff/management relationship is poor.

Among things likely to provoke ill-will and dishonesty is enforced unpaid overtime, withdrawn privileges, pay rises promised but postponed, low staff discounts compared to the trade norm, no legitimate perks such as the chance of cheap, soiled, damaged or out-dated goods, poor staff welfare or facilities.

Conversely, good conditions, fair wages, generous discounts, and privileges, produce good-will and a contented staff that is well worth the cost. Sometimes small touches that cost nothing go a long way. Special thanks when someone puts themselves out or does something beyond the normal call of duty; commendation and praise for something well done or coping well during a specially busy period are usually much appreciated and engender staff loyalty.

Cash bonuses based on profitability can be a valuable good-will builder too. They give everyone a stake in the success of the business and tend to inhibit dishonesty. Anyone stealing is then stealing from his fellow workers as well as the firm, and is likely to incur their displeasure if found out. This can have a more salutary effect than even the threat of prosecution.

Staff selection

While these various measures will go a long way in preventing employee theft, those that are intent on dishonesty will find loopholes in almost any system. Unlike outsiders such as shop-lifters or burglars, they have the time and opportunity to study the system and find its weak points.

If all staff were basically honest there would be no problem and the huge losses that are regularly suffered would be saved. So it all comes back to careful screening of new employees, this really is the biggest single factor in protecting the business from employee theft.

References are increasingly becoming a minefield, and are unreliable in recruiting staff. An excellent reference may be given by a previous employer who just wanted to get rid of the employee, but softened the blow with the reference. Yet this is dangerous, the writer owes a duty of care and could be sued if the reference gave excessive and false praise to an incompetent employee. On the other hand, a bad, yet possibly justified reference could produce a legal action by the employee. With the availability of legal aid, litigation is all too common by those with little to lose, and even if you win you are likely to be much out of pocket.

With general awareness of these twin pitfalls references are usually bland and not very informative, they are best read and ignored. Any specific point of interest could be checked by an 'off the record' phone call to the previous employer. The same applies to giving references. Don't bother unless the employee asks, in which case, say nothing that cannot be backed up by proof.

A recent survey revealed that much the same can be said of curriculum vitaes – they are unreliable. It discovered that one in four CVs were false. Among the details uncovered were bankrupcy, county court judgments, and dismissals for fraud. Qualifications were faked, gaps in careers were covered over, and there were even changes of identity.

While no screening is foolproof much can be learned about the candidate and his views of honesty at the interview. By skilful use of questions that are designed to reveal his attitudes, professional interviewers can often spot a high-risk applicant, or conversely pick one that is least likely to offend. There are various techniques, but most rely on putting the applicant at ease so that he will speak freely and without inhibitions.

Some may resent the nature of the questions asked, or wonder at their relevance, so a disarming remark would be appropriate at the start. This could be to the effect that the firm's policy is to take a friendly interest in all its employees as individuals and not regard them as merely cogs in a wheel. 'So, we would like to know about your interests, hobbies, and family, how you feel about certain matters, and things that are important to you.'

This part of the interview could be conducted in a more relaxed manner after the 'formal' part dealing with personal details and previous experience.

Direct questions are unlikely to produce the required information, but the answer that the candidate thinks is the 'right' one, that is the one he believes the interviewer wants. It would be pointless to ask 'are you honest?' Important questions must therefore be indirect, often devious, and frequently seemingly casual. They should be mixed in with trivial questions which are given greater stress to divert attention from the key ones.

Here is an example to test the applicant's attitude to honesty. A question on his interest in foreign travel could be asked. This could then be followed by another: 'Supposing you came into some unexpected money, let's say you found a wallet with £1000 in it, what sort of holiday travel would you spend it on, which country would you visit?' Or perhaps a candidate who revealed a liking for music could be asked whether he would spend the money on going to live concerts or buying a better hi-fi.

'Red herring' comments made in advance can be used to divert attention from the real purpose of such a question. Thus remarks about enterprise and seizing opportunities could precede the following question: 'Suppose you have just put a few gallons of petrol in your car at a filling station. You then find that the pay booth is empty and there is no sign of an attendant anywhere. Would you be content with your good fortune and drive off, or would you go back and fill up, or would you fill the spare can as well?' (Any reply involving leaving money should be treated as dubious and would at the very least show naïvety, as left money would soon disappear.) The correct thing to do would of course be to leave one's name and address on a piece of paper and call another time to settle up.

As a further example the conversation could be turned to insurance and its value, after which this problem could be posed: 'The tax and insurance on your car have expired but you only have enough money to renew one of them. Which would it be? Why?' Whichever is chosen implies the applicant would be prepared to drive without the other, and without waiting until both could be afforded. The reason given could also be illuminating.

Other questions could be framed in a similar manner to trap those who are not too fussy about honesty. Honest applicants will react unmistakably to them. This may seem to be devious and even unfair, but really, the only way to defeat employee dishonesty is to avoid dishonest employees.

With young candidates, the family background can be a significant factor. A stable family is more likely to produce a regard for principles of honesty than a broken one, although of course there are exceptions.

Any applicant on drugs is likely to be a very high risk, as these suppress moral inhibitions, and the habit requires large sums of money to maintain. If this is suspected, indirect questions as to whether he thinks that certain drugs should be legalized, and whether the authorities are taking too strict a line on drugs, could reveal his attitude. These could be followed by a casual direct rider to clinch matters, 'have you ever tried drugs yourself, perhaps at school?' However, habitual drug takers are usually drop outs and are rarely found among job applicants.

Expensive tastes and lifestyle are also to be suspected. These are unlikely to diminish and have to be supported somehow. If the wages ever prove inadequate, the 'solution' needs little imagination.

People are frequently influenced by their companions, especially the young. Discreet indirect questions as to the applicant's associates and friends may cast a further light on the character of the applicant on the principle of 'birds of a feather. . .'

The manner in which the applicant is attired can also be revealing. Casual dress at an important interview indicates a casual attitude. That attitude could extend to honesty as well as to the firm's business.

No valid judgements can be made that are based on an applicant's race. This is not said to appease the race-relations industry, but happens to be true. All races include some persons who are scrupulously honest, and unfortunately, many who are otherwise. It has been this way at least since the days of ancient Greece, as attested by the legend of Diogenes searching with a lantern in broad daylight for an honest man.

Even those who profess a religious affiliation seem just as liable to immoral or dishonest practices as those who claim none. So this by itself cannot be taken as a positive recommendation. There is though one noteworthy exception. The Jehovah's Witnesses have a strong code of morality and honesty which is scrupulously observed by each baptized member, even in small matters. As a result they have come to be recognized as trustworthy and sought-after employees by many firms. One well-known department store with branches in every major city has a policy of employing them whenever possible. If one turns up for an interview, and is qualified for the job, look no further.

Interviewing suspects

As well as assessing the suitability and trustworthiness of job applicants, skilled interviewing techniques may be required when questioning those suspected of dishonesty, fraud or corruption. While this is not a particularly pleasant task it is sometimes necessary, and is all part of the security scene.

Many such interviews end inconclusively because the interviewer is too trusting, is too ready to accept plausible explanations, or feels that it is rude or impolite to ask too searching questions. There are certain techniques which professional interviewers use to uncover the truth. These are a few:

The interview should be carried out at short notice so depriving a guilty suspect of the time to frame a plausible story; one who is innocent does not need it. A psychological advantage is gained if the suspect is on unfamiliar ground, i.e. in the manager's office rather than on the shop floor. Furthermore, the interviewer should have his back to the window, leaving the suspect to face the light so his reactions can be clearly observed.

The interview should then go through four principal phases. First, and in a polite manner, come questions to elicit the facts surrounding the matter and determine the suspect's version of events. Questions should be asked on related matters that may already be known to the interviewer to test the suspect's veracity. Notes can be made of the substance of answers. The suspect can be asked to repeat or rephrase certain statements. Contradictions and anomalies as well as clues that the suspect is concealing something will often be shown up by this.

Second, in a more severe tone, unsatisfactory or contradictory answers are reviewed and broken down. The seriousness of the offence is stressed plus the certainty that the guilty party will be uncovered. Any evidence of guilt is introduced followed by a direct accusation. The suspect is thus put under stress and guilt reactions begin to show.

Signs of guilt include:

Indirect answers to direct questions.

Unsupportable 'reasons' offered instead of emphatic denials ('I couldn't have carried the T.V. out because I had a bad back').

Protestations of innocence involving phrases such as 'I am telling you the truth', 'truthfully', 'honestly', 'I swear it', 'on my honour'.

Avoidance of pointed, direct words such as 'thief' 'stolen' 'fraud'; guilty persons prefer to use softer euphemisms.

Emotional outbursts such as outrage, tears, or even laughter; although innocent persons sometimes express anger.

Freudian slips such as 'I never went near the till', when nothing had been said that money had been taken from it.

Failure to look the interviewer directly in the eyes, and blinking.
Shallow breathing, and stammering.
Sweating, and touching the face.

By now, a guilty suspect will begin to realize that he has probably given himself away, his position is very weak and that there is little to be gained by continuing the pretence. Now comes the third phase of solicitous sympathy. The interviewer expresses understanding of how easy it is to fall into temptation, it could happen to anyone. He may move around from his formal position behind the desk to be closer to the suspect and reinforce his new role as confidant. He persuasively urges the suspect to get it off his chest, then it all can be sorted out and finished with.

A confession will likely follow, and the final phase is to get all the details and write them down 'so as to clear it all up once and for all'.

The relationship now is usually a friendly one, the offender feels a sense of relief to which the interviewer has contributed. At no time should bullying or bluster be used, the hallowed principle of law remains that a person is innocent until they are proved guilty.

15 Other pitfalls

It may be thought that the various frauds and scams we have already dealt with give a more or less complete picture of the security hazards facing businesses today. Sadly that is not so. It makes one wonder why anyone in their right mind stays in business at all! Here are some more.

Fraudulent invoices

A common fraud is to obtain payment on false invoices. One firm of fraud investigators said that two-thirds of all business fraud were procurement fiddles. These invoices may be submitted by a fraudulent outside firm or injected into the system by dishonest staff. In the former case, such firms submit an invoice for goods never received in the hope that it will not be checked, and often it isn't. In other cases they have supplied overpriced goods that were not ordered. Some years ago there was a big scam operated by some firms supplying carbon paper. Reams of the stuff would arrive, low-grade and highly-priced, enough to fill the company's needs for years. It was assumed that someone must have ordered it, so the invoices were duly passed for payment.

With staff-originated invoices, one tip is to suspect any invoice that has not been folded. All genuine ones are, as they are sent by post and are folded to fit an envelope. Frequently, fake invoices generated by computer are not folded, or may be folded in half instead of into the usual thirds. This is a real give-away.

One firm of fraud investigators runs a computer check on all employees and supplier invoices. In 10 per cent of cases there is a correlation. With some there may be the same bank account, or the same address. One would have thought that employees bent on fraud would have taken steps to conceal these obvious connections, but many do not and so are soon caught.

Some invoices do not have a telephone number, or if they do, no fax number. This should also arouse suspicion. If a number is given, a phone call during working hours would soon reveal who, if anyone, was there. The invoice address could also set alarm bells ringing. A box number, accommodation address or one in a residential area should raise doubts, although if for a small firm, the address could be genuine.

Another possible give-away is when personnel authorized to spend up to a certain amount place a number of orders just short of their ceiling. A further one is when invoices are settled in a few days when the normal is 30 or 60 days.

Company credit cards afford a good combination of delegating purchasing power where needed yet achieving central purchasing control. Many fraudulent transactions have been brought to light by firms that have used them.

Phone hackers

Not the home computer enthusiast, but organized criminals getting into the phone hacking scene, are the ones that are costing businesses tens of thousands of pounds per day, according to one survey. They break into company telephone networks by using the passwords given to employees to listen to their voice-mail messages or make international calls. Powerful computers are used to discover authorization codes in minutes that would take a home hacker months of trial and error. The codes are then sold on the black market, and not just to one purchaser, for they can be sold many times over. Thus a whole retinue of illegal users are adding to the unfortunate company's phone bill.

With an increasing number of firms using voice-mail, the pickings for the criminal fraternity from this source are fat indeed. Examples of the losses sustained by companies are: £500,000 increase in phone bills over several months by one large company with 12 phone lines; a £60,000 increase in four days by another, and believe it or not, £1 million worth of illegal calls were made on the switchboard of Scotland Yard! Some 6 per cent of the top companies admitted they had been targeted, but many said they did not know whether they were victims or not.

So what is the defence? Really there is only one: change the password frequently. Not only will this stop the losses straight away, but the criminals are not likely to try again if they get complaints that the passwords they sell are useless after say, a month.

Telex fraud

This is a new version of the electronic funds transfer fraud. When a request to withdraw funds from a bank by electronic transfer is made, an agreed telex password is used. If the book of codes falls into the wrong hands all sorts of transfers can be authorized, so keep the codes safe.

Company theft

This may seem incredible to anyone who has not encountered it, but one of the latest scams is the theft of a whole company! Not isolated cases either, because they have been reported at running at half-a-dozen a month. How does it work? The fraudsters find a smallish company on the register at Companies House. They then send in bogus changes of the directors and company address. These particulars are not usually checked by Companies House and so are entered as authentic.

The purpose is that now the new 'directors' have a credible identity, credit rating, and a long trading record taken from the firm they have taken over unbeknown to the genuine ones. So they can obtain credit and order goods with no questions asked. Having got as much as they can, they then disappear and leave the mess for others to clear up.

The result is that the company now has a zero credit rating, and the directors may even be liable for the losses. Firms supplying the bogus directors will have lost their goods, and banks, their money.

How can it be prevented? The main responsibility for prevention should lie with Companies House, where all notifications of registration changes should be checked out, if only by phoning the company back. However it seems that at present they are very lax in this. So companies must defend themselves as best they can. The records at Companies House should be checked several times a year. If any documents that should have arrived at a certain time from them have not turned up, inquire why immediately. Any strange communications should not be put aside but investigated straight away.

For prospective suppliers, credit references are not enough. Do not accept large orders by post or the phone from new customers, even if they are a known firm. Visit the premises and make sure that the order actually came from the firm. If this is not possible because of distance, check the firm at Companies House, especially to see if there have been any recent changes. If orders come in from existing customers marked for delivery to an unusual address, phone back to check it. It may be that the firm has re-located, which happens quite often, but it could be a bogus set-up that has taken over the firm's name.

The Nigerian scam

This is a scam that has caught many businesses, and although it has become well-known, it could re-appear in another form. Successful frauds rarely are dormant for long. The victim receives an impressive official looking letter. The letter opens: 'I am (a name is given) the Chief Engineer to the Federal Government of Nigeria Contract Award and Monitoring

Committee in the Nigerian National Petroleum Corporation.' A story follows to the effect that surplus government funds are being held that need to be spent in the UK for the purchase of equipment. However, a bank account that is based in the UK is necessary for the operation. Would the recipient of the letter allow his firm's account to be used? If he is agreeable, a sum of (usually millions of dollars) will be paid in to his account, and he will be awarded a commission of up to 30 per cent for his trouble.

It seems too good to be true, and it is, which all scams are. Once access to the account is given, it is emptied and of course no sum is ever paid in and nothing is ever heard of the 'Chief Engineer' again. The high 'commission' quoted should alert the victim that all is not well, but greed often overcomes caution, and the thought of such high rewards wins out.

It seems that this has now spawned a secondary money-spinner in the operation of this scam. Firms are selling pre-printed letters to criminals for them to use. These make it even more obvious that they are bogus, because the gap left for the sender's name is usually larger than the name used, and the name typed is nearly always in a different type-face from the rest of the letter. Yet, people are still being caught by it.

What action to take

When becoming a victim of any of the frauds mentioned here, the reaction is to report it to the police and let them take over. Certainly the police should be informed, but this may not be much help in recovering the loss. If major, the case will be handed over to the Serious Fraud Office. Reports have suggested that they are not as effective as they could be. They are underfunded, their staff underpaid and are not highly trained. It seems that the best people in the force go into crime detection rather than fraud. Furthermore the main object of the SFO, being a police department, is to secure convictions; recovery of losses is secondary.

The main accountancy firms have highly trained people in their fraud investigation departments who have a good success record in running to earth perpetrators of business fraud. If the sums concerned are large, it is well worth obtaining their services as soon as possible. Then, take civil action against the fraudsters. Civil action requires a lower standard of proof than the criminal courts; it is quicker and more efficient. Cases are on record where such action has been taken and losses recovered, while prosecution in the criminal courts was still awaited. Often too, where a string of frauds have been perpetrated, the courts will allow prosecution of only one or two of them to save juries a long drawn out trial. Thus, given a wily defence, the chances of conviction are limited.

Legal liabilities

Fraud is not the only snare awaiting the unsuspecting business. Increasingly, litigation with the award of substantial damages, is another hazard, and from an ever widening band of litigants and causes.

One is e-mail. Employers are liable for for any e-mail messages sent by employees during the course of their business. A frightening fact! It is commonly assumed that an e-mail is read then wiped and so is not a permanent record. This is not so, it can be made permanent and it can leave traces unknown to the recipient. Thus an unguarded piece of office gossip can be the basis of an action for libel, as the courts regard e-mail as a publication. It is especially damaging if it should concern the reputation of another firm. Recent awards for corporate defamation have ranged from a quarter to half a million pounds.

Libel is not the only e-mail hazard. Litigation could be triggered by e-mails containing trade marks, copyright material, bad professional 'off the cuff' advice, obscenities, racism, sexual harassment, provocation, threats, and much else. A defence that the e-mail was only in jest and not intended to be taken seriously will not be accepted, nor will the excuse that it was intended for one person only.

So what can be done? Clear and uncompromising statements should be circulated and displayed as to what is and what is not acceptable for transmission over the Internet. Warnings should be also displayed that any violations would be liable to instant dismissal. Some form of monitoring should be instituted, and spot checks made on outgoing e-mail.

There are many other sources of litigation which attract large awards for damages. Unfair dismissal, illness from passive smoking, food poisoning in the canteen, undue stress, sex or race discrimination against job applicants, or in job advertisements, harassment, sexual or otherwise, have all been subjects of lawsuits. Breaching copyright and using unlicensed computer programs are others.

It is evident to any reasonable person, that while some of these are justified, many, if not most, are not, but merely try-ons to get money for nothing, which is the prevailing ethos of this age. Unfortunately the courts tend to side with the litigants and many a business has suffered under the unjust burden of having to make a large payout to a totally unworthy recipient.

There is no blanket defence against these; each one must be considered separately and measures taken to guard against them. Being aware of what traps there are is half the battle, as appropriate steps can then be taken. Often this seems to involve wasted time and effort, but it is not if it saves the firm a hefty award for damages.

For example, a most unsuitable job applicant may turn up for an interview. If there is the slightest risk of a charge of race, sex, or other discrimination being made later, you should go through the motions of a full interview and appear to seriously consider the applicant's qualifications. This, even though another applicant has been interviewed that is clearly superior and the obvious choice for the job. When advertising, the apparent net must be cast wide, even though you know you will narrow the qualifications in your final choice. This wastes your time and that of the unqualified applicants, but to do otherwise is to risk a charge of discrimination. You must cover your back at all times.

The disabled

There is an area of anti-discrimination law which is justified, and not difficult to comply with. It concerns the disabled. The Disability Discrimination Act now makes it illegal for employers to discriminate against disabled people in recruitment, promotion, and relocation. It requires employers to make 'reasonable adjustments' to the work environment to enable disabled people to carry on a job without too much difficulty.

This may involve re-arrangement of office furniture, access, and equipment. For many disabled persons, computer operating is a suitable job because it does not involve much physical moving around. Yet there may be difficulties in operating the keyboard due to slow or weak finger action. Keyboards can be obtained that help with these problems. Some have 'sticky keys' which avoids the need to press two keys at once, others have a time lag between the moment a key is pressed and when it starts repeating. There are small keyboards that require shorter hand movements, and voice-operated programs are ideal for many.

The cost of any adaptions or equipment is not usually high, but 80 per cent of any cost over £300 will be met from the Employment Service's Access to Work scheme. The Computability Centre, which is a registered charity, will advise on any specific problem involving disabled persons and computers.

Computer hacking

Data theft, corruption, and viruses cost an estimated £1.2 billion in a recent year. Considerable disruption of normal work and disastrous loss of valuable data can result from the activities of hackers, they are not a minor nuisance, they could put a firm out of business, so defensive measures should be a top priority.

Viruses are short programs that invade the computer's hard disk and destroy or corrupt data stored there. There are very many different ones dreamed up by those with nothing better to do, and like biological viruses that affect humans, they have different effects; some are more serious than others. One of the principal causes of infection is running floppy disks that are of dubious origin. Pirated games are frequently responsible. No disk of unknown origin should ever be loaded into a computer. This should be a strict company rule. If it is not necessary to have floppy drives on work stations, it would be wise to have them disabled.

The other source of infection is directly via the telephone line. Here again it may be asked, is it necessary for all computers on the network to be connected to a phone line? Apart from the risk of hacking and viruses, a connection is an invitation for staff to waste time surfing the Net. Many firms prevent their computers being fitted with a modem, and unless it is necessary to have this facility it is prudent to do likewise.

There are a number of programs that offer protection against hacking and viruses. One is IronWall. It offers 16-bit encryption, stops the computer being used when the user is absent (so preventing unauthorized tampering) and it can lock out communication ports, so preventing the use of a modem. It can disable floppy drives, or allow them to operate only with bona-fide encrypted disks. It will support four passwords that each make available different drives and accessories. Thus four levels of use are possible from the same computer. There are other such software, and new programs are being constantly developed. Investigation as to what is currently available and appropriate to your use is certainly worth the effort.

It may seem that the business man or woman has little time for actually running the business because of the large amount of attention needed to counteract crime in all its forms. Unfortunately this is true to a certain extent. For larger firms it may be an idea to employ an anti-crime manager. Security staff will no doubt already be on the payroll, but the anti-crime manager would have a much larger remit, dealing with all the aspects of security that we have described which would normally be out of the range of ordinary security personnel. He would regularly check all these possibilities, and be alert to pick up on any new ones that the criminal fraternity may devise. Seeing the large sums that can be lost, and even the whole business, this would be a wise investment. For smaller firms, perhaps the various classes of crime could be split up and delegated to existing suitable staff. The thing not to do is to ignore it and think, 'this will not happen to me'. It could be the famous last words.

16 Fire alarm systems

Although the incidence of major fires is far less than that of burglary, shop theft and employee theft, its consequences can be far more serious. There can be loss of life, as well as total loss of stock and buildings. It is therefore something to which very careful consideration should be given. Being very much a threat to security, we here outline what is required for an effective alarm system. It should be noted that most fire stations have a Fire Prevention Officer who is available to advise on all aspects of fire prevention and control. His services are free and should be made use of.

Fire-fighting equipment

The first requirement is to have fire-fighting equipment available; for large premises this means on every floor and on various parts of each floor. The correct type of extinguisher must be used for the most likely type of fire in any given location. Some types of fire and suggested extinguisher are shown in Table 2.

Table 2

Type of fire	*Type of extinguisher*
Carbonaceous, paper, wood, straw, textiles, funishings.	Water, CO_2, soda acid, foam, general purpose powder.
Inflammable liquid oil, petrol, paint tar, paint, spirits.	Foam, powder, BCF, CO_2.
Gas; calor, propane, natural gas.	Powder, BCF.
Metal: sodium, calcium, phosphorus uraniam, plutonium.	Metal fire powders.
Electrical fires, computers.	CO_2, BCF.

CO_2 = carbon dioxide gas; BCF = Bromochlorodifluoromethane, a vaporizing non-toxic liquid giving a clean vapour and no deposits, and so is harmless to electronic equipment and machinery.

Fire blankets that smother a fire are most effective in some cases such as kitchen conflagrations.

Fire risers providing high pressure water supplies for hoses should be available at strategic positions and well marked.

Distinguishable systems

A fire alarm system should be readily distinguishable from an intruder or any other type of alarm. To avoid frequent false alarms, detectors are set to trigger above a certain level of heat or smoke. While some do react in earlier stages of the fire, these are not suitable for all locations. So, a fire can be well established before an alarm is sounded. As minutes, even seconds, can then be vital in saving lives and confining the fire, it is imperative that there be no confusion as to what the alarm means.

While it is technically possible to connect fire detectors to the 24 hour detection circuit of an intruder alarm system, this should never be done except in small domestic systems or in a small shop where the type of emergency is immediately identifiable. Even with these, the battery operated self-contained sensors with their own high-volume sounder, are much better.

To aid rapid identification, fire bells or sounders should have a quite different tone from that of the intruder alarm. Furthermore they are always painted red, while intruder sounders are often finished in many different colours. Thus a stranger, who would not know the difference in sound between fire and intruder alarms, should get the message if the noise was coming from a red bell. The fire alarm control panel is also often coloured red, and unlike the intruder control unit, should be prominently displayed to view.

Detectors

There are several types of detector, each being best suited for a particular purpose. The *optical smoke detector* uses the Tyndal effect whereby light is scattered by particles of smoke. A small radiator directs a pulsed beam of infra-red light at an angle into a detection chamber, while an infra-red detector 'looks' into the chamber at a different angle. As the chamber has a matt black interior, no infra-red radiation is reflected toward the detector, so it sees nothing.

If smoke enters the chamber, the particles scatter the infra-red in all directions and some affect the detector. The first couple of pulses are ignored to avoid false alarms from random dust particles, but any after that triggers an alarm condition (Figure 66).

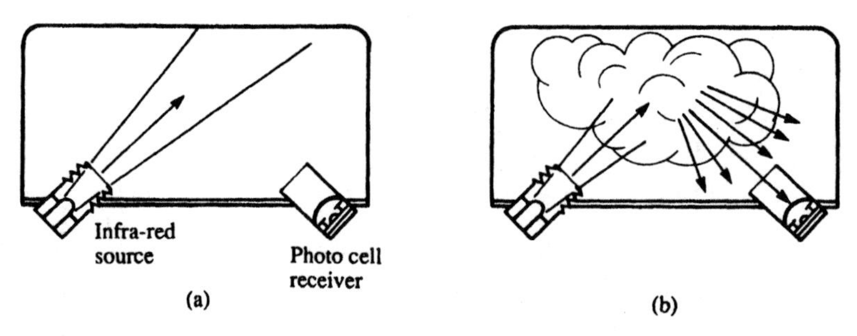

Figure 66 Optical smoke detector, Infra-red source radiates into matt black chamber (a) and the photo-cell receives no reflection. When smoke is present (b), reflections are received by the photo-cell.

Another device that detects smoke is the *ionization smoke detector.* This has a small low-level radioactive radiation source such as Americium 241, which maintains a flow of ions through the air in the detection chamber to a pick-up electrode. Smoke particles entering the chamber pass between the source and electrode thereby inhibiting the ion flow. The electrode current is thus reduced and the drop is sensed by a comparator circuit (Figure 67).

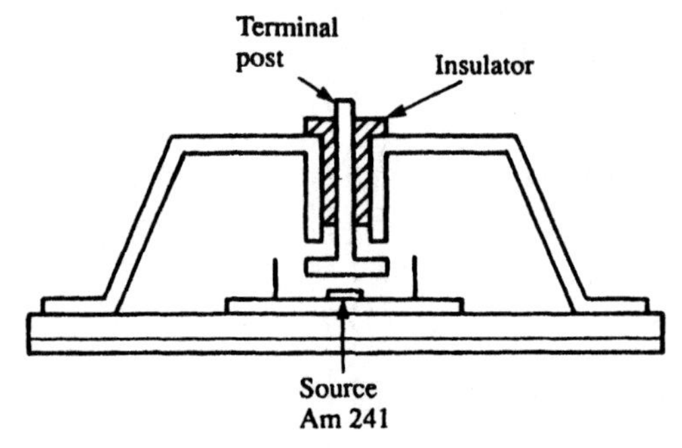

Figure 67 Ionization smoke detector. A radioactive source ionizes the air between it and an electrode causing a current to flow. Smoke particles impede the flow, so reducing the current.

Also relying on smoke detection is the *infra-red beam* which is similar to that used for intruder detection. In this case the transmitter and receiver face each other at a high level where the beam will not be interrupted by humans or objects. A delay in response is incorporated, so that momentary interruption is ignored. Smoke impedes the infra-red beam and when the receiver circuit senses a drop, it signals an alarm.

Other detectors rely on heat, one type being the *fixed temperature heat detector.* The device is normally set to operate at 135° F (57° C), at which temperature the alarm circuit is activated. Another type is the *rate-of-rise heat detector* which triggers when the temperature increases rapidly, typically 40° F (22° C) per minute. The latter usually include a fixed temperature limit as well, so that a slow climb to the set temperature will also trip the alarm.

Heat detectors can be either mechanical or electronic. In the former, a bimetallic strip such as used in most appliance thermostats is employed. This strip is composed of two dissimilar metals bonded together. They expand at a different rate, so that when a rise in temperature causes one to expand more than the other, the strip bends sideways and can actuate electrical contacts. In this case, it releases a latch which tilts a tube of mercury. Contacts are immersed in the mercury so completing a circuit.

With the electronic type, a thermistor, which is a resistor that changes value when heated, is used. Its resistance is compared by a comparator circuit to that of a fixed resistor, and when the difference reaches a certain level, the comparator signals an alarm. In the case of the rate-of-rise detector the thermistor resistance is compared to that of another thermistor which is slower acting. If the temperature rise is gradual, they more or less keep step, but if it is rapid, the fast one leaves the other behind. A resistance difference is thus produced which creates an alarm signal.

At a certain point, the effect of the slow thermistor is reduced by a shunted resistor so that there is very little total change of resistance in the circuit, while the fast one continues to change. Thus a difference occurs above this point so triggering the alarm. The rate-of-rise detector thereby also reacts at the fixed high temperature even if the rise has been slow and so combines the features of both types of heat detector.

Another type is the *flame detector.* These respond to infra-red or ultra-violet radiation given off by the flames. The infra-red type distinguishes between flames and other sources of infra-red such as heaters or persons, by sensing flame flicker. Steady sources are thus ignored. The ultra-violet type responds only to radiation in the 200–270 nm band. Frequencies of this order in solar radiation are absorbed by the ozone layer, so those reaching the earth do not affect these sensors.

Which detectors?

Smoke is usually generated long before there is a major temperature rise, so wherever other factors permit, the smoke sensor should be used rather than a heat detector. The optical detector is best at smouldering fires which generate a lot of visible smoke. These include burning fabrics and furnishings, polyvinyl chloride (PVC), and insulation materials in electrical equipment.

Materials that burn with little smoke, such as gas, spirits and solvents are more quickly detected with flame sensors or ionizing smoke detectors. If both types of material are present, the ionizing detector should be used. Sometimes it may detect a fire of the former group of materials in a very early stage, before dense smoke has been generated, so it is a good choice for a general purpose detector. Either type can protect an area of approximately 1080 ft^2 (100 m^2) around one point depending on layout.

The infra-red beam detector can cover a much larger area as the transmitter and receiver can be spaced up to 300 ft (90 m) apart. Smoke generated up to 20 ft (6 m) on either side of the beam is detected, so the area covered can be 12,000 ft^2 (1080 m^2). It is effective with smoky fires, but less so with clean burning ones. However, air turbulence caused by rising hot air and gasses can produce variations of beam intensity. One model analyses turbulence variation patterns, and initiates an alarm when the typical turbulence frequencies produced by a fire (2–20 Hz) are present. It thus is effective for both clean and smokey fires.

Smoke detectors cannot be used where there may be pollutants in the atmosphere, such as areas with a high level of dust or dirt due to cleaning or machining operations. False alarms could be generated as the detector cannot distinguish these particles from smoke. Furthermore, they can build up deposits inside the detection chamber.

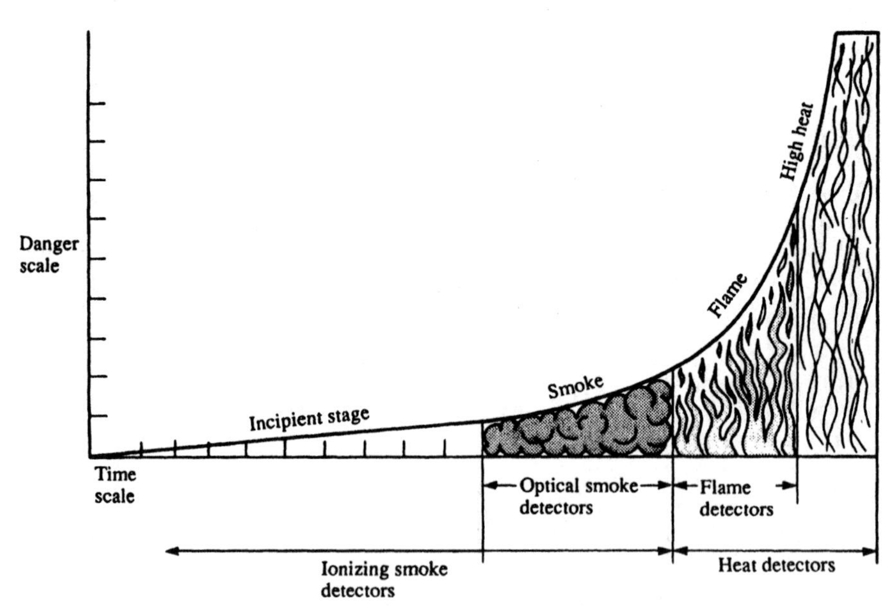

Figure 68 The various stages of a fire and the reaction of different types of detector.

Similarly, areas subject to steam or other vapours such as kitchens, laundries, garages, paint spray shops and bathrooms are likely to prove troublesome. Also cold-storage areas where there is likely to be condensation could produce false alarms. For these, heat detectors must be employed.

In cold and refrigerated areas it may take a long time for the temperature to rise to the alarm trigger level, so the rate-of-rise detector is the one to use. A fire can then be detected at an early stage. Figure 68 shows the four stages of a fire and how the various types of detector respond. Note that no sensor should ever be painted, as this would change its thermal characteristics and its response to heat.

Siting detectors

Having determined the type of detector for a particular location, the number required and the sitings must then be considered. As heat and smoke travel upwards, detectors are fitted to the highest points in the area, which in most cases is the ceiling, When encountering the ceiling, hot air and smoke spread sideways thereby actuating any detector that may not be situated immediately above the conflagration. However, excessive height does reduce efficiency as both heat and smoke can dissipate to some extent on rising through a large area. Heights greater than 30 ft (9 m) may cause problems.

In order to maximize the sideways spread effect at ceiling level, the inlet vents of the detectors should not be below it by more than 6 in (150 mm) for heat detectors, or 2 ft (600 mm) for smoke sensors. This means they should not be mounted on deep girders or beams.

If the roof is pitched or has north lights, the detectors should be fitted in the apex within the above limits. If the pitch is very shallow being less than those limits vertically from apex to base, the ceiling can be regarded as flat and the apex disregarded.

The sideways spread of heat and smoke at ceiling level allows detectors to be spaced out, yet be responsive to a fire anywhere in the protected area. There are, however, maximum distances from any point to the nearest detector, which should not be exceeded if a rapid response is to be achieved. No point should be further than 25 ft (7.5 m) from a smoke detector, or 17 ft (5.3 m) from a heat sensor. For a flat ceiling this means that smoke detectors can be a little under 50 ft (15 m) apart and heat sensors just under 34 ft (10.6 m) apart when in a quadrant configuration (Figure 69).

Each smoke detector thus covers a circular area of 1900 ft^2 (176 m^2) and each heat sensor, 950 ft^2 (88 m^2), However, most areas to be protected are square or rectangular. Inscribed squares in these circles have areas of 1080 ft^2 (100 m^2) and 540 ft^2 (50 m^2) respectively, so these can be taken as the normal area of coverage.

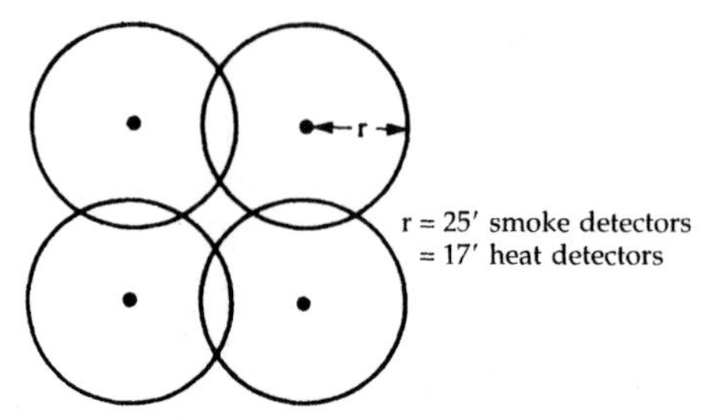

Figure 69 Effective operating radius of smoke and heat detectors, adjacent detectors should therefore be spaced at just under twice this distance.

These figures assume that the ceiling is flat and unobstructed. Any obstructions, such as beams, restrict the sideways spread and so require closer spacing of the detectors. In such case the path around the obstruction should be subtracted from the normal spacing (Figure 70). Thus the total path including that around the obstruction remains the same.

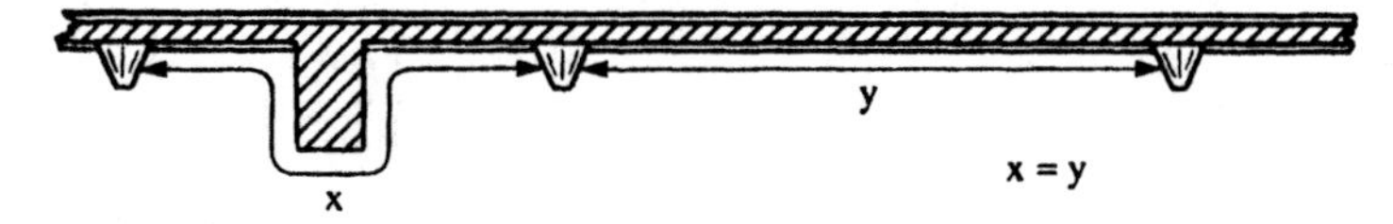

Figure 70 The path around a ceiling obstruction such as a beam should equal the unobstructed path between adjacent sensors.

There is one exception to this rule. If the obstruction has a depth which is greater than one tenth of the floor-to-ceiling height it must be regarded as a wall and the areas on either side as separate rooms. No detector should be mounted closer than 20 in (500 mm) from a wall.

Manual call points

The detectors we have described give automatic protection and raise the alarm both when the premises are vacant and when they are not. But when occupied, someone inadvertently starting a fire or discovering one should be able to sound an immediate alarm.

This is the purpose of manual call points, which are to the fire alarm system, what panic buttons are to the intruder alarm. They consist of a press button which is kept depressed by a glass panel. Breaking the glass releases the button and sounds the alarm. Some panels are scored so that they are easily and cleanly broken by thumb pressure, and have a plastic coating which prevents the operator being cut by jagged edges. Others have a small metal hammer attached by a chain for breaking the glass.

They can of course be broken by an elbow, a brick or any convenient object because the button does not have to be manually operated, just breaking the glass is sufficient. Some have a microswitch that rests against the side of the glass and is released when it is broken. With these, a button is not visible through the glass.

The comparison with the panic button may raise the following questions. Why have to break glass at all? Why not just have the button to press? In the case of the panic button it is located where only authorized staff or house occupants can operate it; they are not likely to raise deliberate false alarms. Fire call points though must be readily accessible for anyone to use, and so are vulnerable to both accidental and deliberate misuse. The glass serves as a deterrent because the noise of it breaking draws immediate attention to that point.

All manual call points should be mounted 4 ft 6 in (1.4 m) above floor level and no-one should be more than 100 ft (30 m) away from the nearest one.

Control unit

Fire control units have similar features to intruder alarm panels, but there are also differences. Like intruder alarms, they can be obtained in single or multizone versions, so that in large premises different areas can be put on different zones, thereby leading to rapid identification of the fire source. Each zone latches on once the circuit has been triggered, and an indicator shows which one until the panel has been reset.

Unlike the alarm control, the panel should be mounted in a readily observable position so that fire-fighters can quickly trace the location of a fire. It also should be located in a place of minimum fire risk so that it will not be quickly disabled by the fire. The main entrance-way will in most cases fulfil these conditions. A card clearly identifying the zones and the areas they cover should be mounted alongside it.

There is no need of a key switch because the panel is left on permanently, but some models have the controls and indicators covered by a lockable glass-panelled door so that they can be seen but not interfered with.

There are no exit problems requiring special circuits, or anti-tamper loops, but fault indicators are commonly found so that warning is given of faulty circuits which can then be quickly traced. An internal buzzer is also used in some models to warn of faults and take the place of the main bells for testing.

Zoning

One zone should serve only one fire compartment, which is defined as any area enclosed by walls, ceiling and floor.

Each storey is a separate compartment unless the total floor area of the whole building is less than 3230 ft^2 (300 m^2) in which case the whole building can be considered one zone.

The floor area of a single zone even if on the same storey, must not exceed 21,500 ft^2 (2000 m^2). Storeys having larger areas than this must be divided into separate zones.

Stairwells, lightwells, liftshafts and the like are single fire compartments even though they extend through several storeys. Each shaft should be on a separate zone.

In cases where dividing walls and split levels may make a compartment difficult to define, the rule is that the search distance necessary to visually locate a fire should not be greater than 100 ft (30 m). It can be seen that these zoning rules not only enable the location, but also the extent of a fire and its confines to be quickly and accurately determined from the indicators on the control panel. This is of invaluable help to fire-fighters in most effectively deploying their resources.

Detector circuits

Detector circuits can be either closed-circuit loops like the detection loops of an intruder alarm system, or open-circuit. With a loop, the detector must go open-circuit when actuated in order to break the loop continuity, and so any wiring break or disconnection is signalled as an alarm. Systems using mechanical heat detectors and manual call points can be connected in a loop circuit and so have the advantage of warning against any circuit disconnection. In the case of a simple open-circuit system, the detector must go short-circuit, and any wiring break is not signalled.

Electronic detectors require a d.c. voltage supply, so an extra pair of wires would normally be necessary to convey it, just as with active intruder detectors, but the number of detectors and wiring used for fire alarms is far greater than that needed for an intruder alarm. So, a four-wire system would considerably increase cable and installation costs.

The alternative which is now commonly used, is the two-wire open-circuit system that supplies the operating voltage along the same wires, and in addition provides continual monitoring for circuit breaks.

In its quiescent state, the detector takes a small current of less than 0.1 mA, but when it is activated, the current increases to some hundred times that amount. It is this sudden surge of increased current that is sensed by the control panel which then triggers an alarm.

The monitoring is achieved by superimposing a small a.c. voltage onto the d.c. The final sensor in the run has connected across it a capacitor and diode which passes part of the a.c. that the control panel continually monitors. Any break in the wiring or disconnection interrupts the a.c. current. The panel then responds by activating a fault indicator. This is better than the closed-loop system in which it is not possible to distinguish between a triggered detector and a circuit break.

Because the detectors draw current, the number that can be connected to any one zone is limited. Assuming a current of 0.1 mA for each, the maximum number is 0.1 divided into the maximum current per zone specified for the panel. Thus a 4 mA maximum zone current allows forty sensors to be connected. Manual call points being only push-buttons, draw no current, so an unlimited number of these can be connected to a zone in addition to the detectors.

Sounders

A larger number of sounders, usually bells are required for a fire alarm than for an intruder alarm. Persons in all parts of the building need to hear clearly, so this means at least one on each floor and often more. The sound level should not fall below 65 dB anywhere. This is just about that of a normal conversation at 3 ft (1 m) distance, and so is not all that loud. If there is any background noise that is likely to persist for more than 30 s, the level should be at least 5 dB higher than the noise.

The rated sound output of a bell is given either at 3 ft (1 m) or 10 ft (3 m). The level drops 6 db for each doubling of distance, or 10 dB for each trebling. So a bell rated at 85 dB at a distance of 10 ft (3 m), would drop to 65 dB at nine times that distance, or 90 ft (27 m). These figures hold good in free air, but would be considerably reduced through doors or around corners. If mounted on a solid structural wall, the sound radiation could be improved.

In locations where there is noisy machinery, extra large mains operated bells can be controlled by the alarm system via a relay, but powered from the same circuit as the machinery. They would then sound only if an alarm occurred while the machinery is running. Otherwise, the standard bells would be heard.

If it is required to wake sleeping persons, such as in hotels, clinics, or rest homes, the sound level must be at least 75 dB at the bedhead. This may be more difficult to achieve because the sound will have to penetrate closed doors as well as having to be three times (10 dB) higher in level. A number of bells may therefore be needed on each floor.

Bell-circuit monitoring

To ensure a high level of reliability, the bell circuits should be continuously monitored just as the detector circuits are. The method used for those would not work for bells, because bells have a much lower resistance than detectors and so would bypass the a.c. monitoring current. The bell monitoring system has a resistor at the end of the circuit and a diode in series with each bell. This passes current only in one direction. When the system is quiescent, a reverse polarity voltage is applied to the bell circuit. Because of the diodes, no current flows through the bells but only through the end-of-circuit resistor. This small current is monitored and if interrupted by a circuit break, the panel signals a fault.

If an alarm is initiated, the polarity of the voltage is automatically switched to normal, and so current flows through the bells thereby sounding them. The resistor being of a high value, has negligible effect across the circuit during an alarm.

Some bells that are intended for fire alarms have a diode included. All diodes result in a voltage drop of about 0.5 V so this slightly reduces the voltage applied to the bell, therefore the output. Fire systems, however, use 24 V or 48 V as distinct from the standard 12 V of the intruder alarm. Higher voltage systems need proportionally less current to obtain the same power than the lower voltage ones. They therefore produce less voltage drop over long cable runs, and the drop that occurs is less in proportion to the original voltage. Thus a 3 V drop is a quarter of a 12 V operating voltage, but only an eighth of 24 V. But with a 24 V system, the lower current would have produced only a 1½ V drop anyway, so the loss is even less.

The only drawback to a 24 V system is that two 12 V standby batteries are required instead of a single one for the 12 V.

Auxiliary functions

The alarm system can do more than just sound an alarm. Most control panels have facilities for switching other equipment when an alarm occurs. Telephone dialling equipment can be triggered, just as with the intruder alarm (see Chapter 6). In this case, of course, the recording for a 999 call would request the fire service instead of the police.

Sprinklers can be activated from an alarm system. These are situated overhead and sprinkle water over a vulnerable area. They do little to extinguish a fire, but can confine it and prevent or slow down its spreading. A sprinkler system needs to be self-contained with detectors in the same area, as inconvenience and even damage could be caused if they were activated by sensors in another part of the premises.

Fire doors can be released by the alarm system. If doors are shut in the affected area, draughts on which a fire feeds are prevented, and the fire itself can be contained, at least for a while.

The door must be fitted with a return spring that closes it firmly. The back of the door is fitted with a catch that engages with a magnetically operated holding device on the wall behind it. This retains it when the door is fully open. A pull releases the door, so that it can be closed manually.

An alarm signal causes the device to release the door and allows it to spring shut. The door can of course be opened by anyone seeking to escape from the area just as in normal use; it is not held in place other than by the spring.

False alarms

The most common causes of false alarms with intruder alarm systems such as movement detectors, unsecured windows and misuse of the exit circuit, do not apply with fire alarms. There is little with these that the user can do to produce a false alarm; most cases are the result of incorrect installation.

Even a loose connection will not generate a false alarm as it does with an intruder detection loop, but instead shows a zone fault. Nearly all false alarms arise from using the wrong type of detector, or siting it in the wrong place. Locations where there are fumes or condensation are likely to produce false alarms with smoke detectors, especially the ionizing type which are more sensitive to the lighter products of combustion.

Regarding locations, a smoke detector on a low ceiling in an area where tobacco is smoked is likely to be triggered. A heat detector would be better here or possibly an optical smoke detector. Heat detectors should be kept away from stoves, cookers, heaters, hot pipes and windows that catch the sun. If these common-sense rules are observed, a false alarm is extremely unlikely.

17 Public liability

An accident to a member of the public occurring on business premises could result in a large claim for damages. While this would be undoubtedly covered by insurance, it could result in much loss of senior staff time, and, if the insurance company was to dispute the claim or be tardy in paying, bad publicity too. Local newspapers thrive on this sort of story. Such incidents constitute yet another threat to the business and its success.

The same applies to employee accidents, especially in factories and workshops where there usually are dangers. But apart from this, no-one wants harm to come to anyone on their premises, whatever the outcome.

The possible hazards are many, and differ according to the layout and disposition of the premises and the displays. We will here list a few of the more common ones.

Steps

These are a very common cause of accidents. An outside step leading to the pavement is particularly dangerous as anyone leaving may be dazzled by the daylight and not see it. If there is sufficient space, the step should be replaced with a ramp, and the longer it is the more gentle will be the slope. This will also permit wheelchairs to have access.

If it is not possible to eliminate the step there should be a clearly defined colour change at the step edge. If the pavement beyond is grey, the step can be black. It should never be the same colour as the pavement. Partially sighted persons in particular rely on such differences to enable them to safely negotiate steps.

During the winter months, the shop will be open during the hours of darkness. There should be a strong light over the door directed downwards so that the step is well illuminated. In the case of a small shop having a narrow shop door which is usually kept closed, a notice can be placed on the inside of the door at head height '*Please Mind the Step*'.

Steps can also be a hazard inside the premises, especially if in a place where they may not be expected. A staircase to another floor is obvious, but two or three steps down to a lower level could easily be overlooked. The situation is worse when the interior is carpeted all over and the steps rendered thus virtually invisible.

The same principles apply here as for the outside steps. First, clearly mark each step. This can be done by fixing white plastic strip along the edge. This should be a wide L-shaped angle that comes down over the edge and so is visible from the lower side as well as the upper. It also avoids the edge lifting thereby possibly causing someone to trip. Next, the steps should be well illuminated, and a notice be prominently displayed. Even with stairways to other floors, the edges should be well marked and the area adequately lit for the sake of the partially sighted. There are incidentally, many more such persons than those totally blind, but unlike the blind they are not easily recognized other than by their special badge depicting a partially-covered eye. Unfortunately not all partially sighted people wear one. Another aid, which is greatly appreciated by the disabled, wherever there are steps, is the provision of handrails. They also could save the unsteady, possible falls, but not all handrails are as helpful as they could be. They should continue for the full length of the stairway without any break even around landings, as the turning to change direction can cause disorientation and giddiness in some persons. For them the continuing handrail can be very reassuring. Rails should not be so wide as to prevent a good grip being obtained, and the supports should be on the inside surface so that the grip does not have to be released at any point.

Obstructions

Any obstruction in the walkways is a hazard and should not be permitted. In some supermarkets, goods trolleys bearing merchandise for restocking shelves are often left unattended. Someone partially sighted or intent on a shopping list could easily walk into one. They often have bases a few inches from the floor frequently with sharp or rough edges or with metal appendages, and the danger to calves and ankles should be quite obvious.

All shelf-stocking should be done before the store opens and goods trolleys should be off the floor before the public enters. If a restocking becomes necessary during the day it should be done quickly, and during a less busy period such as lunchtime if at all possible. In all cases the trolley should never be left unattended.

Boxes, cartons or other such items should likewise be kept out of public walkways and showrooms. Displays should be sited with care, and attention given to any protruding portion. Anything sticking out will sooner or later be walked into or will catch someone's clothing. Especially dangerous are narrow obstructions at face level. Strangely, people seem less likely to notice these than at a lower point, especially if they appear unexpectedly, such as just around a corner.

One 'obstruction' which often causes trouble is an unadorned glass door. Just walking into one, can result in some nasty bruises, but if it should break the consequences can be much worse. The solution to this problem is to stick some pattern on it that harmonizes with the decor, at and just below average eye level. Vertical stripes of suitable colour can look quite pleasing and leave no doubt as to the door's presence.

Falling objects

Light fittings and other fixtures, as well as stock or displays that are above head level are potential accidents waiting to happen, if they are insecure. Special attention is therefore needed to ensure that all overhead objects are safe, and they should be checked from time to time. Suspended objects in particular are prone to descend catastrophically, as cord or wire is subject to fatigue and sooner or later will fail if under tension.

Electrical dangers

Any exposed electrical equipment must be out of reach of any member of the public. The coin-operated electrically-powered children's model motor bikes/cars and similar toys often seen outside supermarkets, amusement parlours and elsewhere, are a serious hazard.

They are no danger in themselves, but the power lead is frequently plugged into an ordinary 13 A switch point at the front of the building near pavement level. If the plug was withdrawn by anyone, especially a child standing on concrete, and a finger touched the live pin the results would almost certainly be fatal. Also the plastic cover could be broken, exposing the live parts of the switch.

Any such models should be connected via a connector which can be locked in place similar to those used for security systems. It should preferably be metal cased to eliminate the possibility of breakage.

Heat hazards

Any unusual source of heat could prove hazardous. Very hot pipes carrying steam or flues, are obvious examples. The public should be kept away from any of these and notices warning of high temperature should be posted nearby. Central heating radiators are familiar to most people, and the temperature although sometimes too hot to keep one's hand on, should not cause a burn from a momentary contact. Other forms of heating, especially sources of radiant heat could be more dangerous and should be well guarded.

Slippery floors

Polished wooden, linotile or plastic floors can be dangerous. Non-slip polishes or other suitable finishes should be employed. Sometimes, otherwise safe surfaces become slippery when wet. Melting snow carried in the shoes of customers can soon transform a floor into a miniature lake. A plentiful supply of buckets and mops or squeegees and people to man them, should be laid on around the entrances as soon as that sort of weather threatens.

Safety officer

The problem often is looking sufficiently ahead, or anticipating possible dangers. After some unfortunate incident it is often remarked 'Who would have thought it?' or 'Why didn't somebody do something about it?' It may be thought that it is in everyone's interests to ensure a safe environment and to avoid hazards, and so is something expected of all. However, what is everyone's job is usually nobody's.

The answer is to make some member of the senior staff responsible for safety, a safety officer in fact. He or she should analyse the hazards, instruct staff on their avoidance, and keep an eagle eye for anything that could pose a danger to customers or staff. Things that need periodic checking, he should check, including the condition of the fire-extinguishers.

Factories and workshops

These are subject to government safety regulations and those appropriate for the particular occupation should be obtained and studied. This is not optional, but is a legal requirement. Mostly they concern the use of dangerous machinery, guards, electrical safety and earthing, the use of corrosive and poisonous materials, the provision of goggles for welding operations, provision of protective clothing, and so on.

One requirement we will briefly consider because it has application in so many different fields, yet one which has been ignored in the past, is that of excessive sound levels. High sound levels can temporarily reduce hearing sensitivity, but the damage can become permanent if the exposure is prolonged. The louder the sound, the shorter the exposure time to produce permanent impairment. The damage is greater if the sound contains impulsive noises caused by percussive elements in the source.

Hearing impairment is centered around the frequency of 4 kHz irrespective of the frequencies and the nature of the sound causing the damage. As damage increases with further exposure, the band of frequencies affected broadens, in some cases reaching down to 1 kHz. This region is the one that most affects the intelligibility of speech, so persons suffering damage may not be aware of general hearing loss but will find it increasingly difficult to comprehend speech although they can hear it.

The maximum permitted sound levels in industry are shown in Table 3.

Table 3

dBA	*Time (hours)*	*dBA*	*Time (minutes)*
90	8	102	30
93	4	105	15
96	2	108	7.5
99	1	111	3.75

Protection against high noise levels can be achieved by wearing ear muffs. It should be noted that employees may not always appreciate the need for these and may at times be inclined to leave them off. It should therefore be made a rigid rule that they should be worn at all times when noisy machinery is being operated.

Conclusion to Part One

The foregoing chapters may seem to have presented a very gloomy picture; so many hazards and risks it is a wonder that any business could ever be a success. Well none of them are imaginary, they are all real enough as the quoted statistics have shown. Of course the amount of risk differs from one place to another and from one business to another. It is up to each reader to assess the situation in his own case and take what steps seem appropriate.

We set out to draw attention to the risks facing businesses today and how they can be either avoided or at least reduced. In this it is hoped we have had success. As the old saw has it, *to be forewarned is to be forearmed*, but this is only true if the warning is heeded. So information needs to be followed up by action, otherwise it is of little value. The first step is to make that assessment to determine just what level of security is required and in which directions. As pointed out earlier, security must be cost-effective, to spend a lot to protect a little is not sound business.

Some of the information in the preceding chapter has been necessarily technical, but we hope that the aim of making it generally understandable has proved successful. We have presented all that need to be known to make informed decisions.

More detailed technical information for those who need it, together with details of practical installation procedures, trade association requirements, and British Standards, are contained in the Technical Section which follows.

Part Two

Technical Section

18 Installing intruder alarms

The system should already have been planned in accordance with the points made in earlier chapters. The plan will specify the type of sensors and where they are to be located. Before installing, a further inspection of the premises should be made to determine where and how cable runs are to be made. There could be more than one possible route and the one chosen should not only be the easiest but also the one affording maximum concealment and protection of the cable.

This installation inspection may reveal that some slight modifications may be required in the original plan, a relocation of the control panel for example. All the cables have to be run back to it, and the proposed position may cause major problems for some of the runs. Having sorted out and noted any amendments, the work can start.

The control panel

The control panel must be fitted securely to the main structure of the building, i.e. not to flimsy partitions or other secondary structures. If mounted on a wooden panel having access to the back, bolts should pass right through with a metal plate at the rear. If there is no such access, No. 10 woodscrews should be used of a length a little short of the wood thickness. If fixed to plastered brick wall, the brick must be drilled and plugged to take screws of at least of 1½ in (40 mm). If the plaster is unusually thick, longer screws should be used. All the fixing holes provided should be used.

The mains supply should NOT be provided via a 13 A outlet socket. It should be taken from a specially installed unswitched fused connector, wired directly to the ring main as a spur. Once the position of the control box has been determined, the supply point is best installed in advance.

The bell

The first step in fitting the bell or any other type of sounder is to drill a hole through the outside wall to take the cable. A long No. 10 masonry drill fitted to a two-speed electric drill running on the slower speed is the best way of doing it (Figure 71). The type of drill having a mechanical speed change is

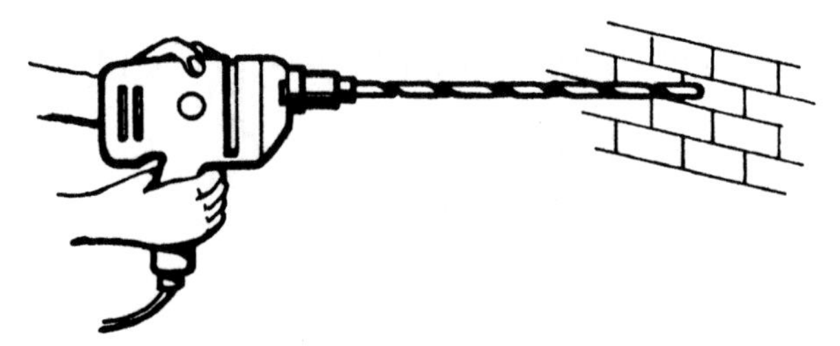

Figure 71 A long Mason Master No. 10 drill is the easiest means of drilling the wall hole for the bell wire.

better than the electronic type, because it develops a high torque at the slow speed, which the electronic speed control does not.

Some walls in older properties are too thick for even a long drill to penetrate. While very long drills are obtainable, these are expensive and unwieldy to use when the length is not required, and are also rather difficult to accommodate in the average tool kit. An alternative is to carry a drill extension. This is a shaft having a female thread at one end to accommodate the male thread on the end of the drill.

As a general rule, it is better to start the hole from the inside as usually precise location of the inside of the hole is more important than that of the outside. Also drilling from the outside could dislodge an area of plaster or otherwise damage the decorations. If the hole is being brought out under the floor, which is recommended, then it would probably be better to drill from the outside, as it may be difficult to drill from between the floorboards and joists. Accurate measurements will have to be made though, to ensure that the hole comes out at the desired place.

Now the bell box base-plate, or the bell itself if of the open variety, can be fixed. Unless the unit is to be mounted on wood such as under the eaves, the wall must be drilled and plugged. The holes should be into the brick or stone, not into the mortar. Rustproofed No. 10 screws at least 1½ in (40 mm) long should be used. An open bell should have a rubber or plastic ring fitted between it and the wall to prevent the ingress of moisture, but with a rough-surfaced wall the gaps should be sealed with a suitable sealer.

Wiring the bell

Ordinary 5 A mains flex can be used for wiring the bell. This is preferable to what is commonly known as 'bell wire' which may have an appreciable resistance over a long run and so result in a voltage drop. If a bell box having an anti-tamper switch is used, four-core cable will be required.

The security of the whole system depends on the bell wiring, so special efforts must be made to protect it over its whole length. It should be run well away from anything which could inflict physical damage. With industrial installations this includes areas of excessive heat.

Where high security is required, the bell wiring should be run in conduit or trunking, but not together with mains wiring. In domestic systems or where lower security is sufficient, the wiring need not be enclosed, but in all cases it should be concealed and if possible buried where it encounters plaster on the down drop to the floor below.

This is not a difficult task; a narrow channel can be cut with a cold chisel and hammer, the wire laid in it and secured with staples, and then it is plastered over (Figure 72). To cause the minimum spoiling of existing decoration, the drop can be run to coincide with a door-frame, then it is only the plaster above the frame that it affects. The cable can be stapled down the side of the architrave for low security systems or buried alongside it where greater security is needed. In most cases only a very narrow channel needs to be chipped out at the side of the wood, and the wiring lodged in it so that there is hardly any noticeable disturbance of the decor.

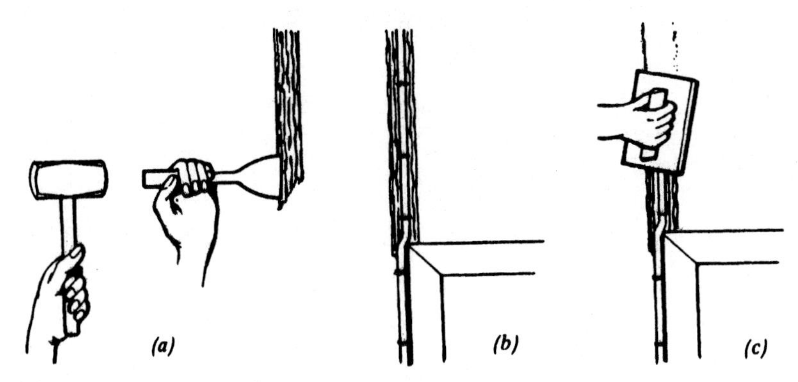

Figure 72 (a) For cable down-drops from the floor above cut a narrow channel in the plaster from ceiling to top of door-frame. (b) Clip cable into channel and continue down side of door-frame to skirting. Door-frame saves having to make a long channel down the plaster, but for maximum security bell wire should be buried in plaster all the way to the control box. (c) Finally, plaster over the wire.

To pass the wire through a ceiling, a small hole can be made from underneath, right into the corner between the wall and ceiling with a long slim screwdriver. A little plaster can be chipped away from the wall so that the hole actually starts below the surface of the wall (Figure 73). Thus when the wire is laid down the channel cut in the plaster there will be no visible sign where the wire is run after the wall has been redecorated.

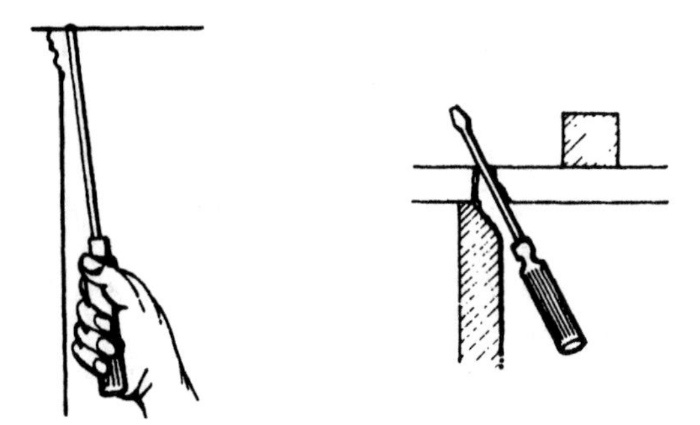

Figure 73 (left) A hole can be made in the ceiling with an old screwdriver. Chip a little plaster from the top of the wall so that the hole is below the level of the plaster. (right) If screwdriver is pushed in up to the handle, the blade will mark the other side of the hole which may otherwise be difficult to locate.

When the ceiling hole is made with the screwdriver, push the tool into the hole up to the handle and leave it there. Then the blade can be seen under the floorboards upstairs, and so the hole is easily located. If the blade is withdrawn it can then be surprisingly difficult to find the hole from above.

The second exterior bell, if fitted, can be installed in the same way, and also the interior one. The bottom part of the bell wire-run where it enters the control box is particularly vulnerable. All wiring entering the box should be buried in plaster and preferably laid in trunking. Fairly wide channels should therefore be cut in the plaster in the direction of the main cable runs, so that all the system's wiring can be laid in as it is brought back. When all wiring is in place, the trunking can be capped, and plastering made good.

Sensor loops

Wiring the loops and installing the sensors is the biggest part of the job. A separate loop is of course needed for each zone. It is recommended that no more than ten sensors be connected to each zone to facilitate rapid identification of an alarm, whether false or actual. Four-core cable is usually required, one pair for the detection loop and the other for the anti-tamper loop. For sensors such as passive infra-red detectors (PIRs) that require power, another pair is needed making six cores; latching sensors need seven, so for them eight-core cable is necessary. For domestic systems not having an anti-tamper loop, two cores are sufficient. Suitable cable is 7/0.2 mm 1 A having a thickness of ⅛ in (3.4 mm).

Most of the sensors will be magnetic reed switches with matching magnets. The switch is fixed to the opening side of the door or window frame, while the magnet is fitted to the door or window itself. Some switches have lead-out wires while others have terminals. Joints to the former should be either soldered and insulated, or made with a junction box, so the ones with terminals are the easiest to install. Wherever possible use the flush mounting variety for higher security. For metal or uPVC windows, surface mounted sensors with leads are required and either soldered or junction-box joints made.

It is easy to make a mistake when connecting multicore cables to sensors, and a single error could take a lot of time tracing and correcting it later. It is a good idea to make a drawing of the sensor terminals and the colour of the wires going to each, then consult it for each sensor connected. The cylindrical flush-mounting terminal-type magnetic switches in common use, have five terminals at the back in a pentagonal configuration, two having a silver finish are the switch terminals, while the three with a brass finish are blank for connecting the other wires. Not all units are of this pattern.

If we assume a four-core cable having red and black cores being used for the detector loop and blue and yellow for the anti-tamper circuit, the five terminals would be connected as follows: blue, yellow and black of the incoming cable connected to the blue, yellow and black of the outgoing, each on blank terminals. The red incoming core is connected to one switch terminal and the red outgoing to the other. This sequence is the same for all except the final sensor at which the blue and yellow are linked (or connected to the same terminal), and the black is linked to the unconnected switch terminal (see Figure 74).

Pressure mats are now usually connected across the detector and anti-tamper circuits, but they have an anti-tamper circuit of their own which is simply a wire that goes into the mat and straight out again. For these, a six terminal junction box is required. Assuming the above colours again, the incoming blue is connected to one of the mat anti-tamper wires, while the

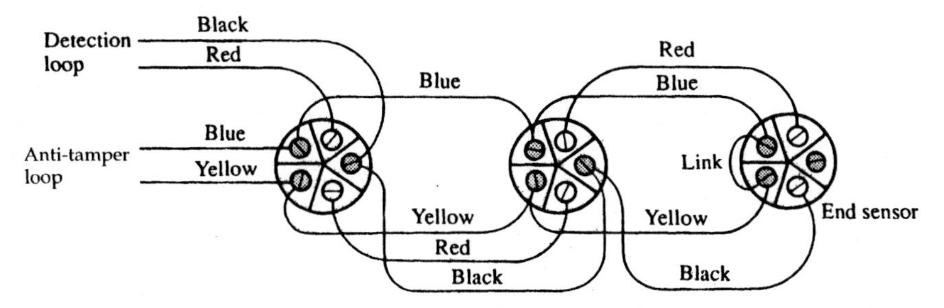

Figure 74 Connection of 5-terminal magnetic sensors. Silver screws are switch terminals, the others are blanks. Wire colours are for illustration only, there is no standard colour coding to preserve security.

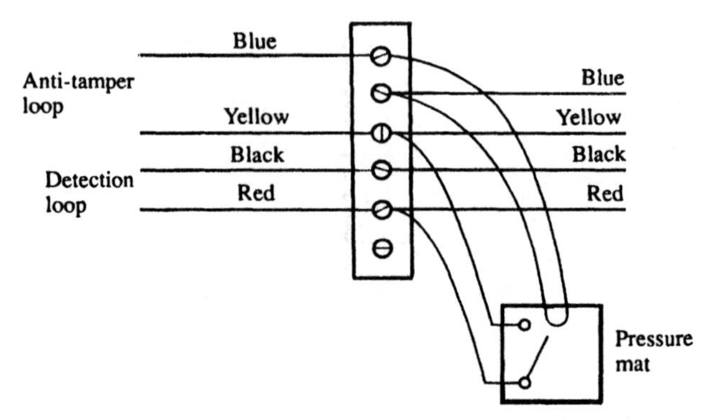

Figure 75 Connecting a pressure mat with a junction box. Connection is across the detection and anti-tamper circuits.

other goes to the outgoing blue. Incoming yellow is connected to the outgoing yellow and also one of the mat switch wires. Incoming black is connected to the outgoing black and to the other mat switch wire. The incoming red is taken to the outgoing red (Figure 75).

The colours quoted and their assignments are not standard but just an illustration to show how the sensors are connected. The four leads from a pressure mat are identified by the insulation which is stripped off the ends of the switch wires but left intact on the anti-tamper ones.

Here is a tip when connecting wires to screw terminals. Wrap the wire in a clockwise direction around the screw, then when the screw is tightened the wire is pulled in toward the centre. If it is wrapped anticlockwise it will tend to be pushed out (Figure 76).

Figure 76 (a) If wires are given a clockwise twist around a screw terminal, tightening will draw the wire in to the centre. (b) If an anticlockwise curl is made, tightening will push the wire out towards the edge where it may come away.

Junction boxes should be made as inconspicuous as possible. Where there is fitted carpet, which is usually the case where pressure mats are used, the box can be screwed to the floor alongside the skirting board and the carpet laid over it.

The result is of course, a lump, but it is less conspicuous than being actually on the skirting, and it could be further concealed by locating it behind furniture. If underfelt is used, a portion could be cut out to accommodate the box thus reducing the height of the protuberance. If it cannot thus be concealed and high security is a requirement, junction boxes can be obtained having a microswitch anti-tamper lid which is connected internally to the anti-tamper circuit.

Running the loops

There are two possible ways of wiring a sensor loop, one by running the wire at floor level along skirting boards, the other by taking it above the ceiling beneath the floorboards of the storey above. The latter may be easier if there is not much furniture and the floorboards are easy to raise. If they had to come up to run the bell wire it may prove most practical to lay in the loop cables at the same time. Loops serving upper and lower floors could also be put in at the same time between ceiling and floorboards. These are the points which can be determined during the previously mentioned installation survey.

Running the wiring along skirting boards causes least disturbance and is probably the most practiced method, but it offers less security than underfloor distribution with which there is no access for would-be tamperers. The degree of security required and the practicability of underfloor wiring will determine which method is followed. For long and complex runs tamper-proof junction boxes should be installed at various points to facilitate future fault-finding.

The main snag with the skirting method is crossing doorways. The wiring either has to somehow pass under the door space, or up and over the architrave which means greater exposure of the cable so reducing security. If there is a threshold draught excluder or carpet joining strip, the cable can be concealed under it. Otherwise it could pass under carpeting providing it was secured at both sides and covered with shallow bevelled channelling (Figure 77).

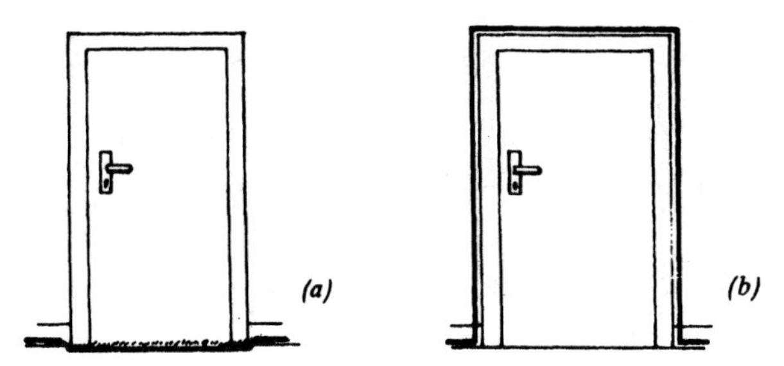

Figure 77 Crossing a door can be problem. (a) If there is a carpet extending through the door or a fixed draught excluder wiring can be taken underneath. (b) If the floor is bare, wiring must be taken over the door-frame.

The location of the sensors on the door or window is determined by which of these two wiring methods are used. If the wiring is taken around the skirting, the contacts will be low, near the bottom of the door, but if the wiring comes down from the ceiling, the sensors will be fitted to the upper part (Figure 78).

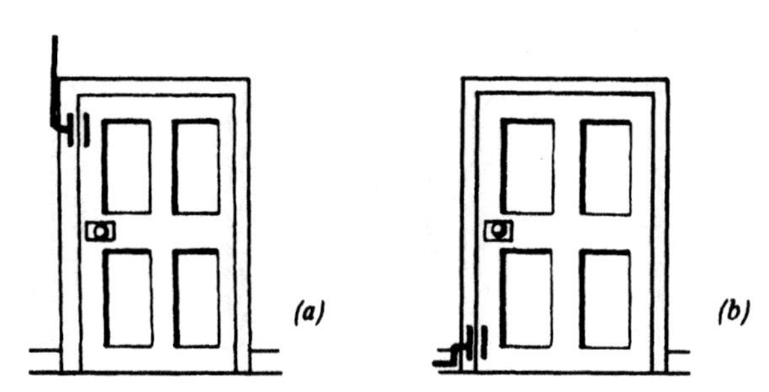

Figure 78 Position of sensors depends on the run of wiring. (a) If dropped from the floor above sensor can be near the top of the door. (b) If run along the skirting, sensor can be low.

Fitting door switches

A ¾ in (19 mm) hole must be drilled in the door frame of sufficient depth to accommodate the body of the switch and also some surplus cable. This enables the unit to be withdrawn if it should need to be checked or replaced at any future time (Figure 79). A neater job results if a shallow cut-out is chiselled to accommodate the flange and so make it truly flush mounting. This is not strictly necessary though as the flange is very thin.

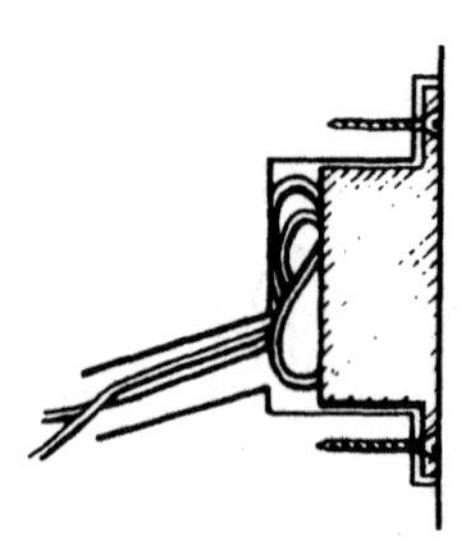

Figure 79 Flush-sensor in position in cavity. Surplus wire needed to enable withdrawal for future examination is stored at bottom of cavity which should be deep enough to accommodate it.

A hole must then be drilled for the cables from the side of the architrave to the bottom of the switch hole. Architraves in many older buildings are so wide that a normal wood drill will not penetrate, especially if the hole must be run on a slant. Other than obtaining extra long drills, two holes must be drilled, one from each end to meet in the middle.

This sounds very difficult, and indeed if drilled in a straight line the chances of them meeting are slim. The chances are considerably improved if both are drilled at a slightly downward angle so that they form a shallow V. This gives a certain tolerance to the vertical angle of the second hole; if drilled at a slightly different angle it will still intersect with the first (Figure 80). The first hole from the cavity can be large to further improve the chance of intersection, as it will be concealed, but the second should only be wide enough to take the cables.

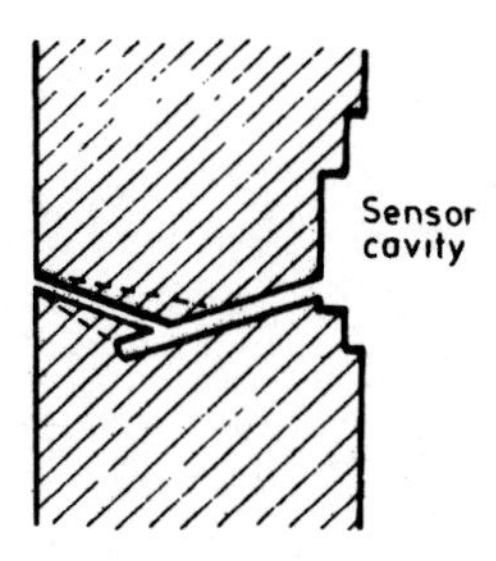

Figure 80 Making two holes meet in a wide door-frame. The first hole from the sensor cavity is made with a large drill at a slightly downward angle. The second hole from the opposite side is of smaller bore, also slightly downward. Dotted lines show the angle tolerance. Hole can be drilled anywhere between these angles and still intersect with the first one.

Having fitted the switch, the mating magnet can be mounted exactly opposite in the door or window. The magnetic field will extend over normal gaps, but excessive space may require special higher-powered magnets, otherwise there may be the possibility of false alarms due to the field being insufficient to hold the switch.

Surface contacts should not be used for business premises other than on metal or uPVC window frames, as they can be defeated by previously attaching a small bar magnet which keeps the contacts closed. This ploy is not detected by the anti-tamper circuit. Surface contacts can be used for domestic systems where such a possibility is remote. They should always be fitted to the inside of the door (Figure 81).

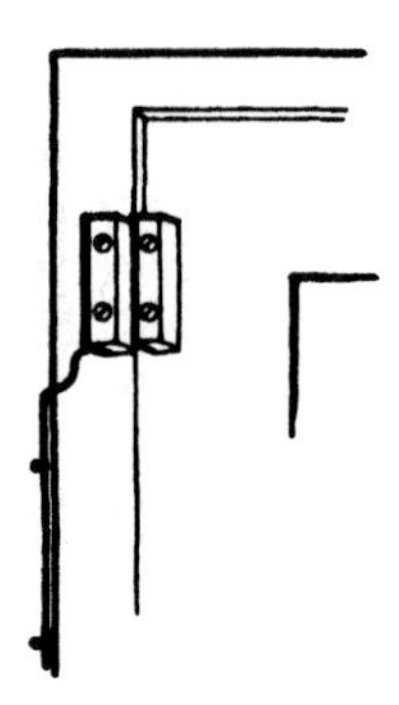

Figure 81 Surface mounted sensors screwed to door and frame.

Window foil

For the highest security each area of glass should have two separate foil runs, one connected to the 24 hour alarm circuit such as used for panic buttons, and the other to the anti-tamper circuit. If either are broken an alarm will be triggered, but it will also sound if the foils are shorted together in an attempt to defeat them. If the control panel does not have a separate 24 hour circuit, a two-pole arrangement can be achieved by using the normal loop plus the anti-tamper circuit. In this case, a short-circuit will not trigger the alarm immediately, but it will when the system is switched on, so warning of its presence then.

An example of a foil arrangement was shown in Figure 19, but any other convenient run can be used that protects the glass without excessively obscuring the display. For highest security, adjacent runs should be no more than 8 in (200 mm) apart. Runs that keep close to the edge of the glass are not very secure, and some coverage of central areas is essential.

The first step is to clean the window with methylated spirit to remove all traces of grease. The window should also be dry and free from condensation when the foil is applied so if the job is done in the winter, some form of portable heating should be placed in the window bay for at least half-an-hour beforehand.

Getting the foil to lie straight is not as easy as it may seem, and a crooked run looks very unprofessional. The solution is to lay masking tape on either side of the path of the run, which will serve a further useful purpose later. It is then comparatively easy to lay the foil in the space between the edges of the masking tape.

Corners are the next problem. Do not attempt to join lengths of foil, the adhesive between them will prevent electrical contact. To turn a corner, first fold the tape in the opposite direction, making a 45° crease. Then fold it back in the required direction and press down (Figure 82).

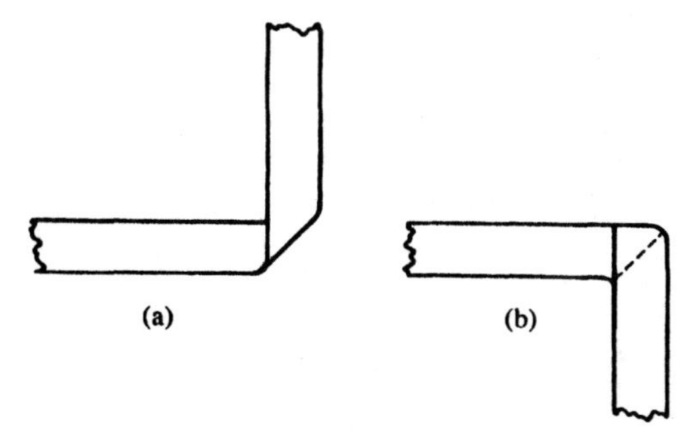

Figure 82 Forming a corner with window foil, (a) first fold the foil in the opposite direction, then (b), fold back to form right-angled corner.

Do not run the foil over any cracks or butt joints in unframed glass, as it will surely break there. Foil is fragile – it has to be to fulfil its purpose, so it needs care in handling. Connections are made by means of adhesive terminal blocks which are fixed to the window glass over the tape, the end of which is brought up and clamped between the block and a metal plate. A screw in the plate connects the wire (Figure 83). Do not leave any loops or raised portions of foil, as these could be broken by window cleaners.

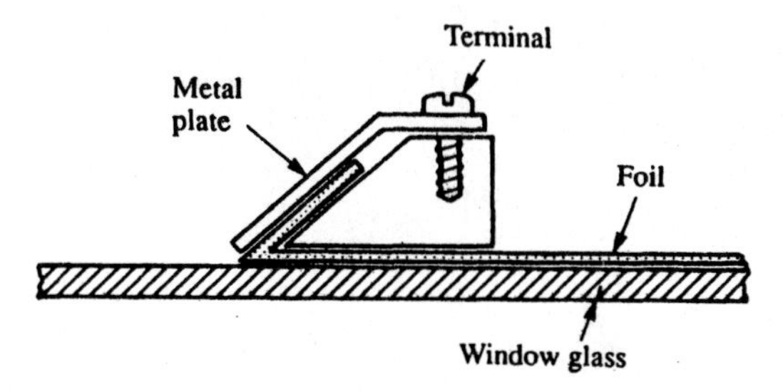

Figure 83 Making off foil with self-adhesive terminal block. Block is fixed to the glass over the foil which bent back on the block and clamped by the terminal plate.

It is preferable, if possible, to mount the connectors near the top of the window as they then are unlikely to become waterlogged by condensation running down the glass. It would also be difficult for them to be tampered with unobserved, if fitted to the top. The worst place is at the bottom. Connectors should be mounted within 4 in (100 mm) of the window frame.

Finally, the foil should be painted with a coat of clear varnish to protect it against corrosion caused by condensation. This is where those strips of masking tape prove their worth, as the varnish extends only a millimetre or so beyond the edge of the tape and ends in a clean straight line. The job is completed by stripping off the masking tape when the varnish is dry.

Vulnerable panels such as thin board partitions can be similarly protected with hard-drawn lacing wire which should be zig-zagged across the area and secured by staples. This wire is supplied on 330 ft (100 m) drums so quite large areas can be economically protected. After fitting, it should be covered with hardboard to prevent accidental breakage because it too is fragile (Figure 84). As it is concealed there is no need of an anti-tamper connection.

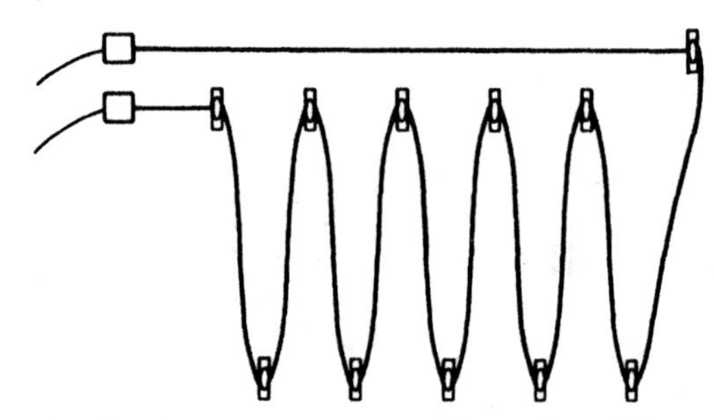

Figure 84 Lacing wire fixed to protect structural member. Adjacent runs should be no more than 4 in (100 mm) apart, and fixing points no more than 2 ft (600 mm) apart.

Vibration, impact, inertia sensors

These should be installed according to the maker's instructions which may differ from model to model. Vibration sensors are the simplest and are just connected in series with a 24-hour detection loop. They must be installed vertically the correct way up, and after installing, the sensitivity must be adjusted by means of a set screw. Where there is a lot of normal vibration such as on main roads or near machinery, the sensitivity must be set low to avoid false alarms. In a more peaceful environment the sensitivity should be set high.

Impact detectors can be mounted in any position. They are powered and so need a six-core cable. Each unit has its own self-contained analyser to provide rejection of incidental impact shocks and set the sensitivity.

Inertia detectors can be either flush or surface mounting. They can be used in window frames at any convenient spot instead of magnetic switches. For fencing they should be fixed at 10 ft (3 m) intervals, while for walls the spacing can be 12–16 ft (3.5–5 m). For the latter they should be mounted away from floors and ceilings and abutting walls. All sensors must be mounted vertically, and an arrow is usually inscribed on the unit to indicate the correct way up. They require an analyser which counts the impulses and their severity, registering an alarm only if these exceed a certain pre-set value.

A number of detectors, up to fifteen with some models, can be connected to each analyser, but where different sensitivities are required for different areas, multiple zone analysers are needed. Each sensor is wired with a four-core cable of which one pair is a series detector loop which is connected to the analyser, and the other the anti-tamper loop that goes back to the control panel. With some models the final sensor has an end-of-line series resistor so that any short-circuit across the loop increases loop current which is detected by the analyser and triggers the alarm. The analyser is connected to a 24-hour detection loop at the control panel, and also to a 12 V power supply. It is best to locate it near the control panel. With fencing, the sensors need to be fitted in electrical conduit boxes if not already suitably enclosed, at the upper part of the fence, and the wiring run along the fence in conduit.

Active infra-red detectors

With active infra-red system, a 12 V supply is needed for both the receiver and the transmitter, but these need not be from the same source. The receiver can be supplied from the control unit in the same cable as the detection loop but it may be more convenient to obtain the power for the transmitter from a small power unit located near it. This could be so if the beam was required to protect a space between two buildings. In such case it would be necessary to ensure that the transmitter was always powered.

An anti-condensation heater is usually included in both units which requires a 12 V supply. Transmitters need 25–100 mA, receivers around 30 mA, and heaters up to 250 mA. The control panel may not have sufficient auxiliary power to provide heater power and a separate supply may be necessary. Separate cables should be used as will be shown later.

If used to protect a perimeter fence, the units should be mounted several feet inside the fence so that the beam could not be cleared by jumping off the top. At least two vertically aligned beams should be used in parallel so that birds flying through one would not cause a false alarm. Alternatively, a double-knock analyser could be fitted which ignores a single short break. While external infra-red units will generally operate in sunlight, it should not be allowed to fall directly on the active surface of the receiver.

The use of mirrors to bend the beam and so cover two or more adjacent sides of an area is frowned upon by some authorities. Each reflection from a clean mirror reduces the beam by some 25 per cent, but there is a much greater reduction from one obscured by condensation. As a mirror has no demisting heater the possibility of dew causing false alarms is a real one.

Passive infra-red detectors

For indoor use, the PIR is superior to the active beam-breaking system. They are now the most commonly used space protection devices. Many installation firms are using them in place of magnetic switches because they mean less work, but this practice is questionable. They allow the intruder to get inside before the alarm is triggered. The perimeter should be the first line of defence, and space protectors should be the back-up or should fill in where it is impractical to fully protect the perimeter. Apart from the light-control PIRs which are independent self-contained units, those used with alarm systems are either latching or non-latching types.

The relay contacts in the non-latching PIR sensor are open-circuited when the device is activated by a moving heat-radiating body and the detection loop is broken, just as with magnetic door contacts. When movement ceases the relay contacts return to their former position. In addition an indicator light comes on, and goes off when the movement stops. This is known as a *walk-test light* because it enables the sensor to be tested for range and coverage by walking within the protected area with the alarm system switched off. (The needed 12 V supply is available from the control panel when it is switched off.)

With latching PIRs, the light stays on after the alarm source has been removed and the panel switched off, thereby identifying which sensor was responsible. This is particularly important when checking out a false alarm, so if more than one PIR is to be used on the same zone they should be of the latching type for this reason. An extra wire is needed to supply a switched 12 V from the control panel when it is switched on. Non-latching PIRs are wired with six-core cable for the detection, anti-tamper, and power supply, while latching models need eight-cores, one of which is unused.

Siting PIRs

Care must be taken in siting the PIR to achieve the desired coverage and also eliminate the risk of false alarms. In the reference section can be seen a number of typical coverage patterns and a sensor should be chosen that gives the nearest pattern to fit the particular situation. In most case the horizontal coverage is that of a right-angle. This suggests that the best position in any rectangular area is in a corner. If mounted midway along a wall there will be blank areas on either side. If such a position is unavoidable, there are models that have a wider coverage. Alternatively, the PIR can be sited on a wall facing the access point. Then the blank areas will not matter as they could not be reached without passing through the active regions.

Range is another factor to be considered. The usual ranges are from 30 ft (9 m) to 50 ft (15 m). For most indoor applications other than large stores, factories and warehouses, this is ample. As nearly all PIRs aim downward the height of the unit governs the range (see Figure 85). A lower position shortens the range, while a high one achieves the maximum; 6–9 ft (2–3 m) is the usual. Most sensors have two or three vertical detection regions to cover distant and near areas, and as shown in the illustration, a high position widens the blind area immediately below the detector. The best height will thus depend on whether the long range or the short is the most important for a given situation.

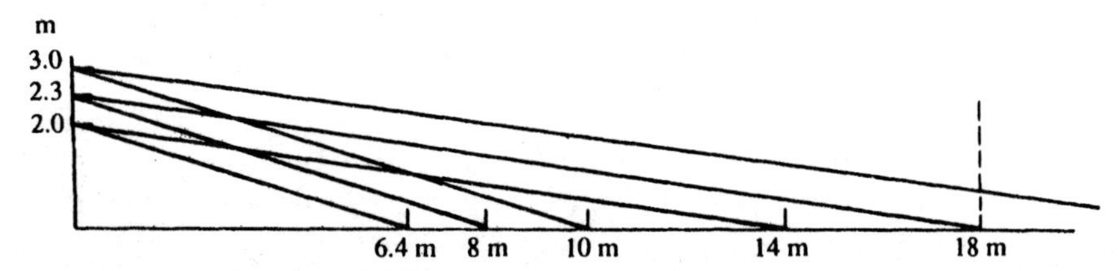

Figure 85 Plot showing how range is affected with height of a PIR at normal mounting angle and also tilted forward 10°.

Some units have an adjustable tilt facility, either of the case or the reflector inside. This enables the desired coverage to be obtained at any height. A high position is preferable to make tampering more difficult, so this facility is a useful one.

While removal of the cover actuates the anti-tamper circuit, a PIR can be easily incapacitated by sticking paper or tape over the lens. If this was done in advance, during normal business hours, and was done neatly it could pass unnoticed and would not be revealed as a fault when switching the system on. Frankly, an anti-tamper circuit for PIRs is rather superfluous as no intended intruder who knew his onions would fiddle with the cover or the wiring when there is such an easy way to defeat the device. This is why they should be mounted above normal reach. It, incidentally, is also the reason why total reliance on PIRs is unwise in business premises. They should only be used to back up perimeter sensors.

Where there is a possibility of objects near the limit of the range causing false alarms, a sensitivity control in some models, permits the range to be reduced.

The detector should not be pointed toward a forced air heating duct as the movement of warm air could trigger it. It is best to site it so that it aims in the opposite direction. Other forms of heating usually cause no problems. It also should not be directed toward a south-facing window. Glass attenuates infra-red radiation, so penetration by body-heat from persons outside would not affect the sensor, but movement of very warm objects heated by the sun could. Apart from these there is little that can create false alarms. Direct sunlight should be prevented from entering the lens as this could damage the sensor.

Among the coverage patterns shown in the reference section, three are worthy of note. One has a totally horizontal distribution and is intended where pets or other small animals could wander into the area. It should be mounted waist-high. In theory it is less secure than other types because a human could crawl underneath the detection region and it is at a level at which it could easily be tampered with. However, the intruder would have to be aware of its presence and pattern which is unlikely. Tampering is a possibility that can only be checked by daily inspection and regular walk-tests.

Another unconventional pattern is the long narrow one. The range is up to 120 ft (36 m) but is confined to a narrow pencil-like pattern with three vertical regions. This is ideal for long corridors and passages.

The third pattern is like a conventional one, but turned on its side. It can be used to protect tall narrow areas such as tall windows and alleys in stock rooms. The range is short at 34 ft (10 m).

Ultrasonic detectors

Being more prone to false alarms than PIRs, ultrasonic detectors are far less popular than they once were. The principal difference in the detection characteristic is that the PIR is most sensitive to sideways movement across the detection field, whereas the ultrasonic device is more sensitive to movement to and fro in it. However, while this is so, both are quite sensitive in the other direction and there is little reason to choose one more than the other on this score.

Wiring is similar to that for the PIR detector, a pair for the 12 V power supply, a pair for the detection loop and a pair for the anti-tamper loop. Some ultrasonic sensors have a latching facility which need an extra wire.

Siting and setting up is very important to minimize the risk of false alarms. Environments where there may be high-pitched sounds such as from steam escaping, gas heating, some types of machinery and brake squeal from passing traffic, must be avoided. The detection pattern is not so clearly defined as that of the PIR so the device must be aimed well away from anything that is likely to move including curtains.

The better ultrasonic sensors include some discrimination against small and random movements such as minor air turbulence and insect flight but they cannot entirely distinguish between different types of movement. A walk-test indicator is provided which flickers on when it detects movement and stays on for a few seconds when the movement is sufficient to cause an alarm.

A sensitivity control is provided, which should be set to the minimum required to produce an alarm when there is movement at the boundary of the area to be protected. Any higher setting will increase the possibility of false alarms.

Exit circuit

If at all possible, only one sensor should be actuated by anyone leaving the premises after switching the system on. In some circumstances, where the control panel has to be situated some distance from the final exit this may not be possible, but the number should be kept to the absolute minimum. Even so, the exit time should not be more than 2 minutes.

Care must be taken in siting space protection devices, such as PIRs in nearby areas to the exit route, that they cannot be actuated by someone on the exit route. It is best that space protectors are not used as the exit sensor.

In most cases the exit sensor will be the magnetic switch fitted to the exit door. The choice here is between wiring it separately to the timed exit circuit, connecting it into the normal detection loop and fitting a shunt switch, or fitting a remote control switch. When wired to the timed exit circuit the normal four-core cable having detection and anti-tamper wires should be run straight back to the control box with no other connection made to it.

A shunt switch can be either a key-operated switch fitted to the exit door or frame if wide enough (Figure 86), or a switch incorporated in a security deadlock. The latter has the advantage of requiring only one key. Lock-switches usually are of the single-pole double-throw (SPDT) type commonly called changeover, and offer either opened or closed contacts when locked.

Figure 86 Key-switch for mounting in exit door.

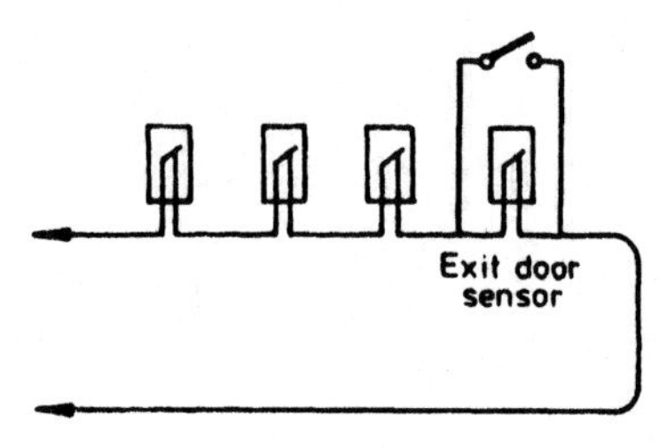

Figure 87 Bypass switch connected across the sensor of the exit door. This can either be a key-switch or lock-switch.

When used as a shunt switch it is connected so that when unlocked, the switch shorts out the sensor (Figure 87). Wiring from the lock should be taken up the back of the door in trunking or metal piping that is well secured to the door, and taken across to the frame by means of a flexible loop (Figure 88). The loop should be terminated in anti-tamper junction boxes; some loops are supplied with boxes already fitted to the ends. The wiring from the magnetic switch is also taken to the box on the frame as is the wiring to the rest of the loop.

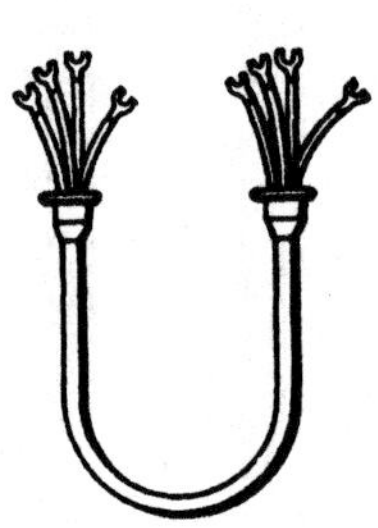

Figure 88 Four-core for connecting door switch to wiring on door-frame. Wires consist of flat wire wound on cord for high flexibility.

A remote controlled system can be switched on and off from the switch lock on the exit door. No operation is therefore required at the control panel itself other than for testing. This is especially useful for public halls which are hired out for various events and the switching on of a security system could often be neglected. In this case, if the door is locked, the alarm system is switched on, so the closed pair of contacts should be used. If a warning buzzer is included with the system, this must be mounted on the door frame so that it can be heard from outside and wired via an extra pair of wires in the cable to the control panel.

Panic buttons

These are normally-closed switches connected to the 24-hour detection loop or if the control panel does not have one, the anti-tamper loop. Although the loop circuit latches electronically in the control box, panic buttons have also mechanical latching which can only be released by means of a key. They do not require a separate anti-tamper circuit, so if run separately from the alarm sensors, two-core cable is sufficient. If on an anti-tamper loop it is advisable to wire them to a separate zone, as then a personal attack can be quickly distinguished from a less-urgent tampering with part of the system.

For small premises the panic button wiring can be on the same run as alarm sensors, but connected in to the anti-tamper loop. In this case four-core wiring will be used the same as the other sensors, but the detection loop wires will be anchored together on one of the panic button spare terminals.

The buttons will be sited according to the specification for the system, which will be wherever a personal attack is possible. At cash checkouts, counter tills, and near front doors and in bedrooms in domestic systems, are the obvious points. They should be easily reached without fumbling, yet not be vulnerable to accidental operation. Some trial positioning with persons likely to use it is worthwhile to get the location exact.

In spite of their large and formidable appearance many panic buttons use a reed switch which has a low current rating. They should therefore not be used to switch current to a sounder direct. Some though are heavy-duty rated and can pass sounder currents; in addition they are SPDT and so can be connected in either a normally-open or normally-closed mode. These can be used in a simple panic system independent of the main alarm, and are wired as normally-open just like an ordinary bell circuit. The mechanical latching ensures that once pressed the alarm keeps sounding. Wiring to the button should be protected or concealed as the alarm would stop if it was damaged.

Outdoor wiring

Outdoor installations are vulnerable to tampering, weather and accidental damage, so these factors must be considered in addition to those applying to inside wiring. If possible, the best way is to bury cable runs at least 1 ft (0.3 m) deep in tough plastic conduit. There should be no joins in the buried cable and the ends of the conduit should be sealed to prevent the ingress of water. If burying is not possible, the wiring must be strung between buildings. It should be as high as possible and supported on a strainer.

On long runs, some consideration must be given to the voltage-loss over the cable. Most multicore cables used for security purposes consist of 7/02

(7 strands of 0.2 mm wire) having a current rating of 1 A. The resistance is 8.2 Ω per 100 m, per wire, so the total resistance for a two-wire circuit is 16.4 Ω for 100 m. This has little effect on a detection loop in which the current drawn is very small.

PIRs take around 20 mA supply power, so from Ohm's Law $V = IR$, the voltage drop over 100 m for a single PIR is 0.33 V. This too can be considered negligible, and up to four could be used on the same cable giving a drop of 1.3 V without ill effect. Most nominal 12 V supplies from control panels tend to be around 13 V. For runs longer than 100 m, and with several PIRs, the drop is proportional and could become sufficient to reduce the sensor's efficiency.

In the case of long-range active infra-red systems, the current can be 100 mA giving a drop of 1.6 V which is acceptable. The demisting heater though can take 250 mA thus dropping a further 4.15 V. Total voltage dropped over a 100 m cable would thus be 5.75 V. However, the heater being thermostatically controlled, switches on and off, so the voltage to the unit could vary by over 4 V as it does so, considerably impairing its performance.

For this reason, the heater should be supplied via a separate pair in the cable and then it will have no effect on the voltage to the unit. Even so, the voltage to the heater will be low, so either a separate heavier twin cable could be run, or a higher voltage supply unit be used to feed it. In this case a supply of 16 V would be necessary. As extra heavy cable is expensive and complicates the run, the auxiliary supply unit is the best solution. The actual voltage required must be calculated from the length of cable, hence its resistance, and the current taken by the heater. It may be more convenient to locate the auxiliary supply at a nearer point than that of the control panel.

Connections

The importance of making sound and lasting connections, especially in the detection loops, cannot be emphasized too much. As there usually are quite a number of connections around a loop, and there are several loops in each system, the chance of sooner or later making a bad one is ever present, human fallibility being what it is. Yet just one poor connection in a whole system can produce false alarms that are very difficult to trace, cause a lot of trouble, and so undermine the effectiveness and credibility of the whole installation. We will examine some of the chief causes of poor connections.

1 A common effect is that many of the strands in the wire bundle are partly cut when the wire is stripped. These can part when the wire is screwed down leaving only one or two strands actually connected. They too can break as the sensor is manipulated into position.

2 Stray wires can escape from the bared bundle and short to the adjacent terminal. This is less likely with sensors having raised insulation around the terminal, but it does happen.
3 An anti-clockwise wire wrap around a screw terminal can force the wire out when tightened (see Figure 76). Even when laid clockwise the wires can slip out from under the screw head during tightening yet appear to be well secured. Sometimes only a single strand or so remains under the head, and this can be broken when the sensor is pushed into place. Contact may be made, but only a touch contact.
4 Another possibility is that part of the insulated wire is included under the head, so, being thicker than the stripped portion, it holds off the head from pressing hard on the bare wire. The result is a loose contact.
5 Terminals in which the screw-end grips the wire, may not suffer from some of the problems of screw-head terminals but they have some of their own. A sharp burr on the end of the screw can sever most of the wire when it is tightened.
6 Also the wire may not penetrate far enough into the barrel of the terminal to be gripped by the screw. A common reason is that the screw was not first unscrewed sufficiently, and so obstructed the entry of the wire. When screwed up tight it may just pinch the end of a strand and so seem to grip the wire. Later, it comes loose.

It can be seen from this list that the chance of making a bad connection is by no means slight, especially when working in poor light, in cramped conditions, and near the end of a tiring day.

The following tips, if rigorously observed, should eliminate the problems.

1 Carry a portable free-standing lamp in the installation kit. It should be powered by rechargeable batteries so it is always ready for use in poor lighting conditions.
2 Always use special wire-strippers, not pliers or side-cutters. There are a number of excellent tools now available that take off the required amount of insulation cleanly, and without damaging the wire. They are also easier and quicker to use.
3 After stripping, twist the bared wired in the same direction as its natural twist before connecting, thereby preventing stray wires from escaping.
4 Use a clockwise wrap under the heads of screw terminals, pulling the wire close to the screw thread, but trim off the excess and avoid laying it over the first part. This creates a high spot that prevents the screw-head gripping the rest of the wire so reducing the contact area.
5 Check that there is a millimetre or so of bare wire exposed outside of the terminal to ensure that no insulation has been trapped under the head.

6 With the barrel type of terminal, look into the end when loosening the screw to see that it is almost fully out and offers no obstruction to the wire. When screwing down, hold the wire gently with the other hand to keep it in place. It is usually possible to feel the wire 'settling' under the pressure of the tightening screw, thus indicating that it is firmly gripped.
7 When the connection is complete give a slight pull to ensure that the wire is fast. Finally when all on the sensor are connected, examine each in turn with a pocket magnifying glass in a good light. It is surprising what potential faults this can reveal and is well worth the extra time spent.

It may be thought that the above is rather elementary and the reader may be inclined to be dismissive. However, the objective is not to reduce the number of bad connections to an acceptably low number per thousand, but to zero, that is the only acceptable figure. More than normal care is the only way of achieving the sort of performance where one *never* makes a bad connection.

Soldering

Some connections in the system may have to be soldered; those to pressure mats in a few cases, and some sensors have stiff wire lead-outs that must be soldered. Making good soldered joints is not such a commonplace skill as may be believed. Even those made by professional electronic engineers have at times been observed to fall far short of the ideal. In view of the importance of sound connections a few pointers here should not come amiss.

First of all the surfaces should be clean and bright. If not, a quick rub with fine emery paper kept for the purpose should do the trick. Wire that has just been stripped can be assumed to be clean. Never use paste fluxes as these can be corrosive; all the flux needed is contained in cores within the solder.

Next, melt a little solder on the bit of the iron, and apply the bit to the terminal wire; the liquid solder offers a large area of contact to the work and so readily conducts heat to it. After a few seconds, solder should be applied to the wire and should be melted by it. If it does not melt, the wire is not hot enough; do not transfer the solder to the bit in order to melt it over the wire, the work itself should always melt the solder.

The solder should run easily over the surface to form a thin coating, it is then said to be 'tinned'. There should be no blobs. The stripped connecting wire should now be similarly tinned and wrapped around the terminal wire. Finally, heat both with the iron and apply a little extra solder which should flow freely over the joint. Remove the iron but do not move the joint until it is set which will be in two or three seconds (Figure 89).

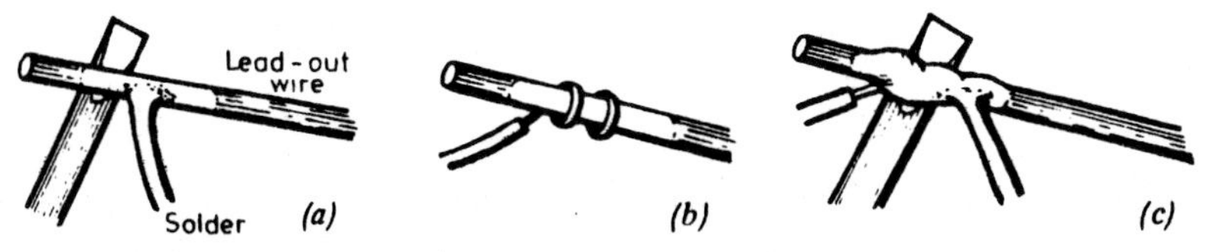

Figure 89 Soldering wire to magnetic reed lead-out wire. (a) Heat the wire with the iron and apply solder to the wire so that it melts and runs easily over the surface, coating it. (b) After similarly coating (tinning) the circuit wire, wrap it around the lead-out wire a couple of times. (c) Heat once more with the iron and apply a little more solder. Solder should flow freely over the joint. Do not move the joint when cooling.

The edge of the joint should taper toward the unsoldered metal, if it curls under, the metal has not been properly tinned and the joint is unsound (Figure 90). The most likely reason for this is an oxidized surface or one having a thin film of grease. Remove all solder and clean with emery before trying again.

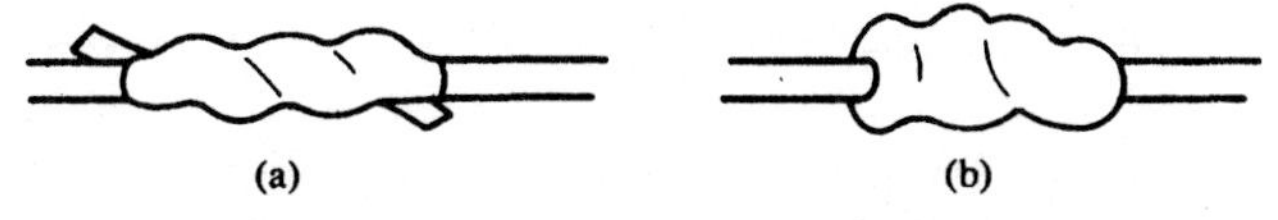

Figure 90 When joined wires are soldered, the solder should flow with a smooth contour to meet the wire at the ends of the joint as in (a); if it forms a blob with thick rounded ends as in (b), the solder has not 'taken' and the joint will give trouble.

When jointing wires from pressure pads, they should be tinned as described, then twisted for several turns before finally soldering. Insulation should be applied to well beyond the joint.

CCTV installation

Installing closed-circuit television (CCTV) is very straightforward. The connections required are a mains supply for each monitor and camera, and a coaxial cable looped from the camera to the first monitor, then from that one to the next and so on. The cable is the same as for television aerials, though for very long runs low-loss should be used. The connectors are of the bnc type which must be soldered to the cable ends.

Looping from one monitor to the next is easy, because each has two sockets that are wired in parallel. So the input cable from the camera is connected to one, while the output is plugged to the other.

In addition, each monitor has a switch which connects a 75 Ω resistor across the cable. Only the last one in the line should have the resistor switched in, which is usually when the switch is up. A terminating load is thus provided which absorbs all the energy at that point and prevents reflections back up the line. If the load is not switched in, ghost images will appear to the right of all objects appearing in the picture.

It may seem that a sharper picture can sometimes be obtained when the resistor is not switched in. This is an illusion, due to a ghost image appearing very close to the original, so outlining or highlighting picture details. Another effect though, is that the synchronizing pulses are distorted, and the picture is liable to suffer from rolling or tearing when the picture content changes. The load should therefore always be switched in on the final monitor, but if it is switched in on any of the others, loss of signal results.

The main consideration is the siting of the cameras. Direct sunlight or a strong reflection should not be allowed to enter the lens at any time, so due allowance must be made for winter conditions when the sun is low. Any bright outdoor scene can be burnt-in to the vidicon pick-up tube if it is allowed to remain, so the camera angle should frequently be changed slightly to avoid it.

Remote control, giving panning, tilting and zooming can be installed if desired. Cable requirements vary between manufacturers, so the instructions supplied with the equipment should be followed. Sequencers require input cables from each camera, usually up to four, and output cable to a pair of monitors. One monitor displays the scene from one camera, while the other shows all four in sequence. The time that each picture is displayed is adjustable from the front panel, as is also the choice of camera for the permanent one. The sequencer thus needs to be mounted where it can be conveniently controlled.

On completion, the system must be set up to give the best results. Aperture and sensitivity controls on the cameras should be adjusted to give a clear non-grainy picture, yet without the 'soot-and-whitewash' effect with flattened highlights obtained when the video circuits are overloaded. If the surveillance is outdoors, take into account the illumination differences at different times of the day. A camera adjusted to give a good picture in bright sunlight will give unviewable results in twilight, whereas one set to give excellent results in poor light may overload in sunlight. Some compromise will be necessary in adjusting the aperture, but auto-sensitivity circuits in the camera will help. A monitor may have to be temporarily placed near the camera, and connected to it with a short cable in place of the permanent one, to enable the camera to be set up.

Monitors similarly need to be adjusted, often surveillance monitors are seen working with very poor pictures that a few moments spent adjusting could greatly improve.

A simple test card can be drawn consisting of a rectangle with a 3:4 proportion and diagonal lines drawn corner to corner. This can be placed in front of a camera and centred using a monitor that is correctly set up, with the rectangle just visible at the edges of the screen. Other monitors can now be adjusted by using it. Height and width controls should be set to fill the screen with the rectangle being just visible. If necessary linearity controls can be adjusted to give correct picture proportions; this is done by adjusting for straight diagonals, if any are bent, the linearity is out. Then, the picture can be centred.

Finally, contrast and brilliance controls are adjusted to give a clear picture. These are often misadjusted resulting in a confusing picture. If all parts of the picture are too bright, the brilliance is too high, but if only the light parts are visible it is too low. Bright whites, but mid and dark greys that are indistinguishable from black, result from too much contrast. A 'washy' picture with grey blacks and pallid whites mean the contrast is too low. Pale colours mean the colour control is too low, while saturated colours mean that it is too high.

A well-contrasted picture has bright whites, dense blacks, and greys that have distinguishable levels from light through mid to dark grey. All this may seem unnecessary for surveillance work, but remember that an identification could depend on how well a face was portrayed on a monitor screen.

19 Testing the intruder alarm, and completion

When all the wiring is finished and the sensors, sounder, and control unit are in place, the last task of the installation is the connection of the various circuits to the control panel. Before this is done, an electrical test of each should be carried out.

A low-reading ohmmeter should be connected across each detection loop in turn and the range adjusted to give as near as possible a mid-scale resistance reading in the case of an analogue meter, or at least one that is well away from either the zero or infinity ends of the scale. This is the most sensitive and accurate region of the meter scale.

The resistance should be read off and noted in a record book and also, possibly a card kept inside the control unit. This could be very useful in tracing future faults.

At present though, the thing to look for is that the reading remains rock steady. It should be closely observed for at least a minute to confirm that this is so. Any slight wavering could indicate a poor joint or a faulty sensor which will almost certainly produce false alarms, and should be investigated before proceeding further. Where junction boxes have been used, parts of the loop can be temporarily shorted out by a crocodile-clip lead in order to localize the section responsible.

Be careful, though, not to be misled by the meter itself. On the low resistance ranges, adjustment potentiometers, internal battery contacts, as well as the test leads can give uncertain readings if slightly dirty. So short the leads together to get a steady zero reading and thus make sure that the fault really is in the circuit.

When a steady reading is obtained, each sensor can be tested in turn by opening the associated door or window. The loop should go open-circuit each time. Next, measure the associated anti-tamper circuit. As this is run together with the detection loop over the same length, the resistance should be similar. However, as the sensor contacts are included in the loop, their presence will make the detection loop resistance slightly higher. The resistance of reed contacts as used for magnetic switches is from 100–250 mΩ (0.1–0.25 Ω), so four switches should read less than 1 Ω, and up to eight, no more than 2 Ω.

If the detection loop has a resistance difference of much more than this, an abnormally high resistance in one of the sensors is possible. This too could be a source of future false alarms and the sensor responsible must be traced and replaced. If the anti-tamper circuit has a higher resistance than the detector loop, there must be a short-circuit in the latter, possibly due to an incorrect connection at one of the sensor terminals.

Measurement should now be made with the meter set to the highest ohms range across the detector and anti-tamper circuits. The reading should be infinity. If pressure mats are wired to the zone, a weight such as a chair or toolbox can be placed on each in turn, whereupon a low-resistance reading should be obtained.

Finally, a reading should be taken on the high-resistance range between the detector loop and earth, and between the anti-tamper circuit and earth. These should also show infinity. Any measurable leakage could set up an earth loop which in some circumstances could trigger the alarm if it falls below a certain value.

Each zone should be dealt with in the same way, and also the exit circuit. The latter will in most cases consist of a single sensor on the exit door. If a shunt switch is included, the action of this can be tested too. With the door shut a closed-circuit should be registered, whether the switch is locked or unlocked. When the door is open, the loop is open-circuit when the switch is locked, but short-circuited when the switch is unlocked. Note that passive infra-red detectors (PIRs) and inertia analysers may have a series resistor which will affect the reading when the circuit resistance is measured.

Unused detection circuits must be bridged at the control panel terminals. Many have links already in place so they have to be removed as the circuits are connected. Unused auxiliary power supplies are of course not linked. Connections to telephone dialling equipment should NOT be made at this point. Measure power supply current and record the reading. Date the standby battery.

PIRs

When the power is first applied to a PIR sensor from the control panel, it takes several minutes to settle into a stable operating mode. This does not apply at each system switch-on because power from the panel is constantly supplied even when the panel is switched off. So the PIR is always ready for action as soon as the system is turned on.

For the initial test though, the warm-up period must be observed. Then with the system switched off, a walk-test can be conducted over the area to be protected, the indicator showing when the sensor is triggered. There are two points to check: the first is that the detector senses movement over the whole area, or at least the vulnerable parts where an intrusion could occur;

the second is that it is not triggered by movement outside the area, especially if such movement could normally be expected when the alarm system is on. Adjustment of the angle and/or sensitivity may be needed to produce the required detection and rejection pattern.

Inertia sensors

A special tool is available for testing inertia sensors. It is a calibrated spring-loaded device which is pushed against various parts of the structure containing the sensors, and delivers a sharp shock of repeatable intensity. The analyser which latches on, has an indicator so that tests can be made and checked by the indicator before the device is connected to the main system.

The sensitivity of the analyser needs to be adjusted so that it is not tripped by normal vibrations, but only those that could result from attempts to break in or through the material. Much depends on the nature of the material. Dense and heavy materials have a large inertia and so do not vibrate as freely as those that are less dense, and so need a high sensitivity setting. Materials that vibrate easily need a much lower setting. Wood or plasterboard requires a setting from 75 per cent to 100 per cent; metal, glass or plastic, 50 per cent or less.

Window foil testing

The most conclusive method of testing window foil is to smash the window. As it is possible that the owner may view this test with a certain lack of enthusiasm, it can instead be assumed that if the electrical test showed an unwavering low-resistance reading, the circuit is in order.

Panel testing

With all circuits connected to the panel, except any telephone dialling equipment, the control panel itself with the detection circuits can be tested. As there will be a number of sensor activation, the main sounders should be muted at this stage. How this is done depends on the type of control panel. Some enable an internal buzzer to be activated in place of the bell by internal switching or linking. Failing this a buzzer or lamp should be connected in place of the bell wires.

The detection circuit of each zone should be tested by actuating one of the sensors. There is no need to operate every sensor as these will already have been checked in the electrical test. Disconnect one anti-tamper wire in each zone to test whether the correct tamper indication is given. Then short one anti-tamper terminal to one detection terminal in each zone; an alarm condition should result.

If the exit circuit is to be timed, determine how much time is required, and set the time on the panel. Then check it. Check the actual exit and entry modes.

Finally, connect the bells and trigger an alarm to hear that both (or more) are sounding. There is no harm in letting them ring for a minute or so at a reasonable time during the day as it will inform the neighbourhood that a working alarm system has been installed. Any other facilities can be tested according to the procedure described in the maker's literature.

Telephone equipment

Automatic diallers and communicators should not be connected to the system until at least a week, but preferably longer, has elapsed from the installation date, to allow any teething troubles to be sorted out. If any false alarms have occurred, connection should not be made until the cause has been traced and eliminated, and after a further period of uneventful system operation.

Testing by means of a simulated alarm can then be carried out after connection and after first advising the recipient, whether a private number or a security depot, that such a test is about to be made. A further call can be made afterward (or they can ring back) to confirm that the alarm message was correctly received. Tests should not be made then or later without prior arrangement as they could be treated as an actual alarm.

Resetting

Some panels reset automatically after a full alarm and the bell timer has expired, so that any further triggering will produce another alarm, provided the detection circuit is not still in the alarm condition. If it is, resetting is inhibited, or in some models all zones are reset except those that are still open. Others are manually reset while yet others intended for use with 999 telephone diallers can only be reset by an engineer. These prevent switching on after a false alarm until investigation by an engineer. Some reduction in security is thereby inevitable.

20 Fault-finding in intruder alarms and maintenance

The most common fault encountered with alarm systems is the false alarm. The causes in order of probability are:

1. misuse by user;
2. incorrect siting or setting of space detectors;
3. excessive vibration, wind, storm or other unusual physical actuation of sensors;
4. wiring faults;
5. poor alignment of magnetic switches;
6. control panel faults.

Not to be overlooked is the possibility of a thwarted attempt at entry which has left little trace. For example, a hefty shoulder charge against an outside door could move the top of it sufficiently to actuate its sensor, but if the lock held, it would spring back into place. The alarm may be triggered but no trace of the attempt remains.

The design of the detection circuits is such as to minimize this sort of thing happening because they ignore momentary breaks in the detection circuit of between 0.2–0.8 seconds. However, if a panel was designed for the faster response, such an event could trigger it.

When investigating an apparent false alarm, the first thing is to check which zone was triggered. Most multizone panels have a latching indicator to show this and it eliminates at least half if not more of the system from suspicion. If the affected zone has sensors on outside doors or windows, carefully examine for any attempts at entry. Look for such things as fresh indentations in woodwork caused by a jemmy or bruised paintwork, scratches, scuff marks or dirt on window sills caused by shoes, or footprints in soft earth.

If an outside door on the circuit has inside bolts as well as a lock, were they bolted? A door bolted top and bottom cannot be sprung inward away from the frame sufficiently to affect a magnetic sensor, but an unbolted one could. Thus eliminate possible outside attack before assuming a false alarm.

Misuse or change of use

An example of misuse could be switching on the system by staff who were leaving while someone was still in the building. Perhaps all zones were activated when one should not have been.

Many false alarms are generated by misuse of the exit arrangements. With a timed exit, someone could take too long to leave after the panel is activated. Or, perhaps they remember something that had to be collected in an adjacent area and so deviate from the exit path, thereby triggering another sensor.

These may not be serious as they are obvious and the persons concerned can in most cases deal with it there and then, switching off and re-arming the system. If the installation is connected to a telephone dialler though, a full alarm will be notified to the number dialled. An engineer may also have to be called out to reset the system.

Another staff-induced alarm could be caused by an object such as a chair placed on a pressure mat. If light, it may not have an effect immediately, but after an hour or so may compress the foam sufficiently to make contact. If any mats are connected to the affected circuit, check that all are well free of obstructions, and question staff whether anything has been moved since the alarm. The offending object might have been put aside thereby destroying the chance of discovering the cause of the alarm.

Perhaps the most common of this type of false alarm is caused by failure to secure windows or doors. They may be shut, but if not properly secured may be moved by wind or vibration. An example of one actual case demonstrates the problems this can cause.

A church hall was wired with magnetic switches on some interior doors as well as the usual perimeter ones. One door which was sprung to keep it closed, led to an ante-room which contained an access hatch to gas-fired heating equipment. The equipment was ventilated by a grill to the outside. Over a period, false alarms occurred that could not be traced, in addition to a real break-in. Eventually the cause was by chance actually observed. When the wind was strong and in a certain direction, it blew through the outside grill and through the hatch into the ante-room. If the door was closed though not properly latched, it blew open, but being sprung, then closed with sufficient force to latch it. Being shut and latched it therefore attracted no suspicion. The door catch was found to be poorly fitting so that it needed some pressure to latch, and consequently was frequently left unlatched. Attention to the latch eliminated the false alarms.

Catches, latches and locks on wired doors and windows should be checked at the time of installation, but they could deteriorate afterward. Alterations such as fitting draught excluders may be made, which could cause a poor fit and the possibility of a door springing open. All those on the affected zone should therefore be examined. Users tend to be careful with exterior doors which of course they should, but are often less so with interior doors.

False alarms and PIRs

The passive infra-red (PIR) detector is normally reliable and not prone to generating false alarms, as are ultrasonic and microwave devices. This is because it responds only to objects that both move and radiate infra-red. If properly sited there should be no trouble, but some possible alarm sources might have been overlooked. A grandfather clock having a pendulum that reflects the suns rays from a window at a certain time of day is an example.

Usually heating systems cause no trouble, except forced air heating, but it is possible that a strong draught of warm air getting into a cold storeroom from a heated part of the premises, could be seen by the sensor as a moving warm object.

Where there are pets or resident mousers, a false alarm is possible unless the sensor is carefully sited to avoid them. An animal could have been introduced into the premises after the system was installed, or a large rodent such as a rat could be responsible.

Where two or more PIRs are connected to the same zone, they should be latching, so the one causing the alarm can be identified. If this is not the case, false alarms can be much more difficult to trace.

If there is no stimulus the PIR cannot trigger, even if set to it's highest sensitivity, however, if an inexplicable false alarm has been experienced, sensitivity should be checked by means of a walk-test to see if it is just sufficient to trigger with a human moving at the limit of the protected area.

Wind and storm

Stormy conditions often produce a spate of false alarms, but do not too readily assume that it was a false alarm. Burglars frequently choose windy nights for their activities because any noise they may make is masked by that of the gale.

There are several ways whereby a storm could actuate an alarm system. Inertia and vibration sensors are perhaps the most vulnerable as vibrations and movements of building fabrics and fencing are likely to considerably exceed the normal. There is little that can be done to prevent this, as reducing sensitivity will only make them too insensitive to adequately protect against intrusion. If a violent storm has been forecast and all the inertia sensors are on one zone, it could be left off while the rest of the system is on. This of course would mean some loss of security, but it would have to be balanced against the trouble and inconvenience caused by a false alarm. Much would also depend on the degree of back-up cover given by the other types of sensor.

Exterior windows and doors fitted with magnetic switches should not be affected by wind unless actually blown open. They should be physically secure enough to withstand rattling that is of sufficient violence to set off the alarm. If not, then the physical security needs improving. However, interior doors with poor latches could be blown open as the example described earlier illustrates. All interior doors that are wired, should have positive latching action. When investigating, do not overlook the possibility that the door might have been closed or blown shut after initiating the alarm.

Water could cause an alarm, if it saturated a sensor terminal block or terminal box, as it could cause a leakage between the detector and anti-tamper circuits. This is more likely if it contains dissolved salts leached from building materials which increase the conductivity of water. Sensors and junction boxes should be checked and dried out wherever there is any sign of water entering the building. Domestic systems lacking an anti-tamper circuit should be unaffected.

Wiring faults

If the greatest care has been taken with the connecting of detection loops as stressed in the previous chapter, there should be no reason to suspect bad connections. However, the system might have been installed by another installer who was not so meticulous.

Another possible cause, is that arising from work by builders, decorators, electricians, plumbers or just someone hammering a nail in a wall to hang a picture. Cables can be damaged in such a way as to cause intermittent faults. This may not show up immediately, so enquiry should be made as to any work carried out in the premises for at least the last month.

Other changes such as furniture re-arrangement, installation of new equipment, clearing out, and even spring-cleaning, are all possibilities that could cause cable damage and should be considered.

After checking out the possibilities mentioned in the previous sections, a cable fault could then be suspected. The test described in the previous chapter could now be made. After disconnecting the loop from the panel, a low-reading ohmmeter is connected across it and a close observation made for at least a minute or so. Any tendency towards changing resistance or wavering, clearly shows a loose connection or defective reed switch.

Location of the fault can be speeded up by selecting a sensor or junction box about halfway along the cable run, disconnecting it, and measuring on from there. Thus half the loop is eliminated, and the faulty half can be halved again, so narrowing the chances down until the fault is found.

Cable breaks

In some cases the loop may be permanently open-circuit, which is certainly easier to trace. Here again there is a method of testing that greatly speeds things up. The instrument to use is a direct-reading capacitance meter. These instruments are now quite inexpensive, will read from a few pF up to thousands of μF and are very useful for equipment servicing.

As damage, more often than not, affects only one or two wires in the cable, the method assumes that at least one wire in the multicore cable is intact. Disconnect the cable from the panel, and also the furthest sensor. Measure the capacitance from each wire in turn to earth. Two capacitance values will be found, some cables measuring one value and the others reading a lower value. The higher capacitance cables are those that are intact, while the lower ones, those that are broken.

If the ratio between the two values is calculated, this also gives the ratio of the distance between the panel and the break to that of the whole run. So if, for example, one value is 300 pF and the other 900 pF, the break is a third of the distance along the cable.

Should all cables measure the same, then all are broken, which can be verified from resistance readings that show both anti-tamper and detection loops open-circuit. If this is so, measure the capacitance back from the furthest sensor. Add this reading to that from the control panel to give the total, and the ratio of the control panel reading to this gives the distance as before.

If there should be a break in the bell wire, the capacitance reading should be taken across the cable from one wire to the other, as there is no easy access to the bell end to disconnect it. A measurement can be taken of a few feet of similar cable and then the distance to the break can be calculated.

When repairing broken cable it is best to rewire the affected section, but if this is not practical, either a junction box should be fitted, or soldered joints made.

Misaligned magnetic switches

This is really due to bad installation, but the trouble may not surface for some while, so its true nature may not be suspected. If the switch and its magnet are not aligned when the door is closed, the magnetic field acting on the reed is smaller than it should be. It may be just sufficient to hold the reed switch on, but there is no tolerance for further field reduction.

All magnets slowly lose some of their magnetism, also the gap between frame and door can increase due to wood shrinkage, or the door may drop or warp, thereby reducing the magnet's influence. So, the switch may function for a while, but its operation may become increasingly critical, sometimes closing and other times not. The result can be perplexing false alarms.

Not to be overlooked is the possibility that the switch and magnet were properly installed originally, but later, a new door was fitted, and the magnet was then incorrectly aligned by the fitter.

Any suspected magnet/switch combination can be tested by actual operation using the control panel internal buzzer. It should be possible to open a door at least an inch before the alarm triggers. If it is actuated by a much smaller door movement, then the set-up is too critical and very likely to cause false alarms. If the alignment appears right and the door gap is not excessive, the magnet may be weak or the reed too insensitive to magnetic fields. Replacement will prove which.

Control panel and power supply faults

These are the least likely units to cause false alarms. By far the majority of faults are those already discussed. However, they do sometimes misbehave. Testing is quite easy: remove all detection, anti-tamper wiring for the affected zone and bridge their terminals. Silence the external sounders and disconnect any telephone connections, then switch on and operate the appropriate zone switch.

Movement detectors can be triggered by a change of supply voltage which in turn can be produced by mains voltage variation or a change in load due to some other device on the auxiliary supply circuit. This will not happen if the circuit voltage is regulated. A test of its regulation should therefore be made by measuring the voltage, shunting a 50 Ω load resistor across it then remeasuring. The difference between the readings should be very small.

After a false alarm

Some authorities recommend that the system not be switched on again after an unexplained false alarm until a thorough test can be made by an engineer. This is why some control panels have engineer-only reset. The reasoning is that the cause may still be present and so produce a further false alarm later.

Frankly, the value of this advice is very dubious. Some burglars are known to deliberately set off an alarm system without obvious entry, then wait for the alarm to be switched off and the fuss to die down before then breaking in without risk of interruption.

How could an alarm be set off externally? One possible way is by using a magnet. Remember that the reed switches are held on by the magnets in the doors or windows; they are released by removal of the magnetic field. If a bar magnet producing a similar field is held against the door parallel with the one inside it, but with magnetic polarity reversed, it could cancel or divert its field sufficiently to release the reed switch.

The position of the sensor can be easily determined by running a pocket compass around the frame to detect the presence of the door magnet. The magnet strengths can then be matched if the applied magnet is stronger by gradually moving it away, turning it to achieve polarity cancellation. Some skill and patience may be required but it is possible. A more likely method is a hefty shoulder charge to momentarily displace the door magnet.

Rather than switching the entire system off after an apparently false alarm there are several preferred options. The simplest is to cancel the zone that originated the alarm, but leave all the others on. This provides some security at least and the intruders will likely trip sensors on another zone. If an engineer is present he can disconnect the telephone equipment if fitted, and/or the outside sounder, leaving the internal bell operative. This option may be preferred, if cancelling the affected zone would leave a large security gap. The internal bell would still serve as a powerful deterrent, yet would cause minimal disturbance to the neighbourhood.

Records

It is desirable, especially with larger installations, that a record be kept of all events relating to the alarm system. This would include a description of the layout, which sensors were on what zones, the location of concealed wiring and resistance readings across each loop. The latter can in addition, be kept on a card inside the control box for quick reference by an engineer. If capacitance tests are made, these too could be noted down. The date of installation would start the record, and any addition or modification should be noted, together with the date.

All alarms, whether false or real, should be recorded, and in the case of the false ones, the cause if discovered, and the steps taken to eliminate the trouble. If an unfortunate history of false alarms should develop, such a record could be very useful.

It may not be necessary to keep a record with small systems such as domestic or small shop premises, especially if not professionally installed. If so installed though, a record is advisable even with these, as it could provide the facts and figures necessary should there be cause for complaint against the installing company. The record should be kept safe and secure as it contains details that could be of use to a potential intruder.

Maintenance

Regular maintenance is advisable for all security systems, whether a simple owner's check of a domestic system, or a professional engineer's test of a large industrial installation. In the latter case, maintenance is essential.

For domestic or small shop premises, the owner will be aware of any accident or damage that could affect the system and will take steps to rectify it as soon as possible. However, a regular routine check should also be made.

The outside bell is the most vulnerable of all the parts of the system, not only to interference, but also from the effects of air pollution. Acidity in industrial areas and salt near the coast can cause corrosion, while other pollutants can gum up the works if allowed to accumulate. It is a good practice, then, to sound the bell by triggering the system once a week. This will confirm that the bell is working, and keep the moving parts free. If it is done at the same day and time on each occasion, and then allowed to run for only some ten seconds, neighbours will not confuse it with the real thing, and it should not create a nuisance.

If a different sensor is used to trigger the system each time, this will test them all in the space of a few weeks. If standby dry batteries are used, a test every two months with the mains supply off, will ensure that they are serviceable. Contrary to what may be expected, this will not deplete them if the test is only for a short time. Dry batteries deteriorate most quickly when they are not used at all; small periodic discharges actually extend their life.

Given small regular discharges, the life of the batteries should extend more than two years, but to be sure it is probably best to change them then.

Professional maintenance

Weekly or even monthly checks are impractical for the larger system, especially if it is the subject of a maintenance contract with a security company. One reason is that a time-consuming detailed examination and testing must be made each time to ensure full security, whereas the weekly check by the owner of a small system is simple and carried out in a few minutes. Even so, routine maintenance should not be less frequent than every six months. Where high security is required, it should be more often.

A suggested routine for a maintenance visit is as follows:

1 Check the system record book for any events since the last call that may require special attention, and enquire if there have been any problems.

2 Following the description in the record book, make a visual check of all sensors, visible junction boxes, and surface wiring. Note any damage, and loose or abraded wiring.
3 Check the fastenings of all protected doors and windows, that all external ones are lockable and have rack bolts where appropriate, and that internal ones have a solid latching action. Observe whether any fresh security hazard has been introduced, such as a new unwired window or door.
4 See that all pressure mats are free of obstruction and that no object has been placed so close as to be a false-alarm hazard.
5 Check all PIRs, that they are still positioned correctly, that there is no change in heating or other factor that could introduce a false-alarm risk. If the walk-test indicator is continually enabled as most are, walk test the perimeter of the protected area to check that the sensitivity is adequate.
6 Examine the control panel, check that it is securely fixed to the wall and that there is no obvious damage. Check that no controls are loose.
7 Mute or disconnect the external sounders and any telephone equipment, leaving only the internal test buzzer, then switch each zone on in turn leaving a sensor open to check that the correct fault registration occurs. Next operate every sensor as far as is practical on each zone, to ensure that each triggers the alarm. Check the exit circuit and its timing. Inertia sensors should be tested with a calibrated shock tool.
8 Remove and refit the covers from sensors and all accessible junction boxes in turn to check that they trigger the anti-tamper circuit. Any zone not having such boxes or sensors should have its anti-tamper loop disconnected to provide the same test. Check also the control panel anti-tamper circuit.
9 Operate panic buttons with the system switched off.
10 Make a physical examination of external sounders, their housings and the security of their fixings. Remove covers if boxed, to check the anti-tamper circuit. Refit. Physically check the internal sounders.
11 Reconnect the sounders and initiate an alarm. Check that all are sounding and that the volume is adequate. (If a sound-level meter is available levels should be measured at a specified position and the result noted in the record book.) Check that beacons are working.
12 Examine the mains supply, that it is powered from a fused unswitched socket, check that fuses are of correct value and examine the ends. (Some fuses are found with poor soldered ends that result in arcing when a load is applied.)
13 Check the date of standby dry batteries. Connect a 25 Ω wire-wound resistor across the battery and measure the voltage. If the battery has been in service for 75 per cent or more of its shelf life, or the voltage on load is below 10.5 V it should be replaced and the date inscribed on the

case and in the record book. Measure and record the current in all switch positions plus that taken by the sounders.

14 Examine rechargeable standby batteries and check the voltage as in 13. Clean any corrosion from the terminals and grease them.

15 With the system switched on, remove the mains supply and then reconnect. This tests that the changeover to battery and back again functions smoothly without initiating an alarm.

16 Finally, telephone signalling equipment should be tested. Tape mechanisms should be checked for smooth operation and the heads cleaned. Recorded tapes can be heard via a handset. Power supplies and standby batteries should be tested as in 12 and 13 if separate, although as most equipment receives power from the control panel the tests will have already been performed. Anti-tamper circuits should be checked. Advise by telephone the security control station, 999 control, or the private number, whichever is appropriate, that a test is about to be made. Connect the line to the equipment and actuate an alarm. Call back to see if the signal was received. (Note than in some areas test calls to 999 control are not permitted or there are strict regulations governing them. These must be complied with.) Leave the unit set to 'manual reset'.

It is obvious that a full maintenance check covering all the above points is a major job, involving some disruption of normal activities in the premises, and so one that is not usually undertaken more often than every 6 months. Yet some points should be checked more frequently, while others being unlikely to change over a short period, could be tested less often. There would thus seem to be a case for copying the motor trade in having a 'short service', with less frequent 'long services'.

A suggested short service could include points 1, 2, 3, 4, 5, 7 (check one sensor only in each zone), 9, 11 and 13. These could be carried out at 2-monthly intervals.

In view of the possible disruption caused by maintenance visits, they should be made only by prior arrangement, and the supervisor should be advised on arrival and again when the tests are completed.

21 The National Supervisory Council for Intruder Alarms

The NSCIA is a technical organization concerned that intruder alarm systems are installed and maintained to the appropriate British Standards. It maintains a roll of approved installers who issue a certificate of status and competence for all systems they install.

NSCIA inspectors will examine, if required, any installation which is the subject of a complaint, and report to the user what remedial work is necessary to bring it up to standard, or if the complaint is unfounded. Arbitration is confined to technical details and the organization will not intervene in financial or contractual matters.

They do, however, offer guidelines as to the obligations and liabilities of both the customer and the installing company, and suggest terms and conditions of rental and maintenance contracts as well as those of outright sales. Sample forms are offered for system handover and maintenance reports.

A code of practice covers these various points and the technical requirements which should be observed by an approved installer. A copy of this 50 page publication can be obtained by any firm engaged in the installation of intruder alarms whether having approved status or not. It is available on payment of the publication cost from:

NSCIA,
St. Ives House,
St. Ives Road,
Maidenhead SL6 1RD.

Most of the technical points made in the code have been covered in earlier chapters of this book. This is a summary of the main ones. Comments are in brackets and italics.

Staff and equipment

Staff employed by the installation company should be thoroughly vetted and all records of installations kept secure. Identification cards with photographs and signature should be carried.

Adequate staffing governed by the number of installations serviced should be maintained; it is suggested that up to 150 requires one engineer; 150–500, two engineers, and one extra engineer for every 250 additional installations, but there should always be at least one standby engineer available. Size of installations, their geographical spread, and contractual frequency of maintenance visits, also influence the staffing level.

Emergency service should be available 24 hours a day, and an engineer should make the call within 4 hours, but at least before the system is due to be switched on.

All engineers should be equipped with hand tools, torch, electric drill and ladder. They should also carry service manuals and data, test equipment and spares. A reliable communication system between the engineers and the controlling office should be maintained.

Adequate training of new engineers should be undertaken, and up-dating on latest models and development given as required.

System design

The nature and degree of risk should be considered in the design, extent and layout of the system. Operational routine, seasonal stock fluctuations, possible future expansion of premises, are all factors to be taken into account. A pre-installation inspection should be conducted to determine circuit runs and anticipate any possible problems.

Reliability is an overriding factor in the design and installation of systems; also important are ease of operation, with accessibility and sufficient test points to facilitate fault tracing.

Sufficient detection zones should be provided, but with not more than ten non-indicating sensors or 1600 ft^2 (150 m^2) of lacing wire (*rather ambiguously termed 'continuous wiring'*) in each circuit. Each zone should have a latching indicator which remains on after the system has been switched off and until it is reset. Where telephone equipment is connected, reset should be by engineer only, but exceptionally, it could be by a subscriber's staff member sufficiently qualified to identify and rectify the cause of alarm.

Latching anti-tamper circuits with indicators, should be included for each zone.

The signalling device chosen should be appropriate for the degree of risk involved and security needed, whether audible alarm plus digital communicator, 999 dialler, or central security station link, or audible alarm only. (*The type and number of audible alarms, and the use of beacons should also be considered.*)

Exit routes should be the shortest possible with the minimum number of sensors encountered en route. Movement detectors are discouraged, and timed exits should not exceed 2 min. Sensors on other zones should be unaffected, and a visual or audible indication of time-period elapse should be provided. If telephone equipment is connected, timed entry should be initiated by a switch on the first entry door or other means, but shunt switches are not acceptable. (*The reason for this stricture is not given, possibly because without the urgency of a timed entry, staff could forget to switch the system off and so generate a false alarm. No objection is raised to shunt switches on non-telephone systems, and indeed they are a very convenient means of entry.*)

Installation and wiring

All parts and wiring should be physically protected against accidental damage or tampering by the public or the subscriber's employees. It should also present a neat and professional appearance. Underground cables should be suitably protected, and voltage drops on long runs should be considered. Mains wiring should be installed with regard to current electricity regulations, and never, without special precautions, in the same conduit as detection loops. No wiring should pass through lift shafts.

Circuit wiring should be identified to facilitate future fault location. (*Identification will be by colour coding, but it is wise in the interests of security not to have a standard code for all installations; coding should be varied, and the code used noted in the record book.*)

Junction boxes should be fitted in complex circuits to enable circuit faults to be quickly traced. At the installer's and insurer's discretion these should be of the anti-tamper type in high-risk situations. (*Junction boxes invite tampering. As anti-tamper boxes are only a few pence dearer than normal ones, they should always be used.*) Cables run in conduit should be protected by bushes or grommets at both ends. Conduit or trunking should terminate as close as possible to the unit it serves. Open wiring can be secured by cable ties or insulated staples, but runs should be along the sides and top of architraves, and in the lip of a skirting board or picture rail, to avoid damaging the face of woodwork by staples. Staple guns should be used with caution to avoid damage to cable insulation. (*Picture rails are best avoided for wiring runs as they are often removed for redecoration.*)

Overhead catenary wires should be securely fixed to the buildings, and the cable supported by loops or plastic buckles. Conveyance of mains wiring by this method must comply with the current electricity regulations.

Installing sensors

Doors and windows should be checked for good physical fastenings, and with insufficient movement when closed, to operate the sensor. Any failing these requirements should not be wired but reported to the owner. Switches and magnets should be concealed as far as possible, yet be capable of being dismantled for service and replacement.

Pressure mats should only be used on smooth, flat, and firm surfaces, and covered with a material that protects against damage, wear and tear. At the time of fitting it should not be possible to distinguish the outline of the mat. (*A new mat nearly always shows a slightly raised profile except on the thickest carpet. It soon beds down to be almost imperceptible.*)

Vibration detectors should not be used on materials liable to expansion and contraction or on semi-rigid structures. Assessment of ambient vibration of traffic, machinery, or structures accessible to the public, should be made and the sensitivity set accordingly. They should only be used on materials with good mechanical transmission, and with analysers. All fixings must be secure, especially when conduit is used to improve efficiency. (*Intertia detectors, which are not mentioned in the code are generally better and less prone to false alarms than vibration detectors.*)

Acoustic breaking-glass detectors are susceptible to existing cracks, impacts not resulting in breakage, high-frequency sounds from brakes, steam or air jets, so should not be used where these are present. (*More sophisticated detectors respond only when low-frequency impact sound is followed by high-frequency breakage sounds and so are virtually immune from these.*)

Contact acoustic detectors should be connected to a system only after a period of measurement of ambient noise within the frequency range and setting the sensitivity accordingly. At least 7 days false-alarm-free operation should be logged before connecting to a system which includes telephone signalling equipment.

Airborne-sound acoustic detectors should be used with signal analysers and set to respond to sounds 20 dB above peak ambient noise.

Lacing wire (*termed 'continuous wiring'*) must be protected from physical damage, especially where it is less than 6 ft (2 m) above floor level. When battens or tubes are used to support it, they should be firmly secured to the protected structure at not more than 3 ft (1 m) intervals, but not anchored to any part of the main structure. That part of the wire leading to the main structure should be formed into loops which are well secured by means other than staples, and permit no more than 2 in (50 mm) of movement before breaking.

Foil on glass should be covered with a moisture repellant coating, and there should be no loops of foil to the connectors, which should be at the highest part of the glass.

Panic-buttons (*termed 'deliberately operated devices'*) should be sited for operation with the minimum of obvious movement, and to avoid accidental operation. If fixed on movable structures, flexible connections suitably terminated should be employed.

Use of movement sensors

These must be sited on vibration-free mountings with an unobstructed view of the protected area, hence the stacking and possible arrangement changes of stored goods are to be considered. They must also be unaffected by movement on the exit route or at any associated controls. Furthermore, care must be taken to identify possible spurious sources of motion likely to produce false alarms. Sensitivity should be set to the minimum for the required range.

Each detector, whether separate or in a multihead system should have its own latching indicator which shows the first sensor to be triggered until the system is reset. A walk-test facility should be included, but it should not be possible to switch the system on when a detector is in the walk-test mode.

Active infra-red beam receiver faces should be aimed to avoid direct sunlight at all times of the year, a snout being fitted if this proves difficult. The possibility of avoidance by bridging or tunnelling should be considered, and the span should be less than the specified range to allow for dirt and dew. Mirrors should not be used to bend the beam.

Parallel beams should only be used where there is a possibility of interruption by birds or animals. Receivers should be electrically in parallel and have auto-reset. Beam axes should not be more than 20 in (500 mm) apart, and it should not be possible to switch on the system if one or both beams are interrupted.

With microwave sensors, the perimeter should comprise metal, concrete or 225 mm brick, most other materials are not an effective screen and movements outside could be detected. Particular hazards are timber doors, asbestos roofing and liquid moving in plastic pipes. A check on possible interference from fluorescent lamps should be made, and also a pre-commissioning measurement of ambient noise (*spurious r.f. background*). It should be 20 dB lower than the threshold signal required to initiate an alarm, unless the detector has analysing circuits to minimize noise response. Two or more microwave detectors should not interact to cause either false alarms or insensitivity.

Ultrasonic detectors should not be used where there are high frequency sound sources, or air turbulence, unless signal processing is included to minimize their effect.

Passive infra-red sensors must be carefully positioned, not directed at any rapidly changing heat source, objects that reflect sunlight, or any hanging objects that could be affected by draughts. (*These would only be a hazard if of a different temperature to the environment.*) Humidities in excess of 90 per cent and temperatures above 86° F (30° C) are unsuitable. The possibility of r.f. interference should be considered before and after an installation. (*Specified temperature ranges of 32° F–122° F (0° C–50° C) are now common, and r.f. interference suppression is usually built-in.*)

External sounders

These should be sited to give maximum prominence, sound output, and protection from wilful or accidental damage, yet afford reasonable access for servicing. The cable should likewise be fully protected.

Minimum fixing requirements are: for a brick wall, three No. 10 screws into suitable plugs, penetrating the actual brick to a depth of at least 40 mm (1½ in). For metal, wood, or thin skinned structures, use bolts with a backplate.

The sounder should not generate electrical interference and is recommended to operate with a timer to cut out within 30 min, in conjunction with an optional flashing light. (*Most panels now have built-in bell timers.*)

Internal sounders

The position should be chosen to generate maximum noise in the protected area and afford full protection for the cable.

Telephone signalling equipment

All equipment should be housed in attack-resistant cases, be firmly secured, be located within the protected area, be fully anti-tamper protected and concealed if possible. The telephone line should preferably be ex-directory, dedicated, and restricted to outgoing calls. The British Telecom (BT) connector should be adjacent to the equipment and connecting cables should be physically and electrically protected.

An integral rechargeable battery with sufficient capacity for five alarm signals should be included, with an automatic charger capable of fully charging it within 24 h.

A digital communicator should respond within 1 s to any alarm trigger exceeding 200 ms, and once started be impossible to stop. It should recognize that contact has been established, and if not, make at least three attempts to do so. Any system connected to telephone signalling equipment should be engineer reset.

Facilities to delay an external sounder may be included, but if attempts to establish telephone contact are unsuccessful within three minutes, the external sounder should sound immediately. Furthermore, the line should be 24 h monitored for incoming calls or a line voltage less than 35 V, and if either condition exists during an alarm situation, the sounder should sound within 30 s. (*The advisability of delaying the external sounder is dubious. It may lead to catching the thieves red-handed, but it could also expose stock and premises to further damage or loss.*)

Power supplies

All power supply units should be sited in dry, well-ventilated positions not subject to extremes of temperature. They should be readily accessible to allow them to be examined, cleaned, tested and measured at each maintenance visit.

Primary cells should be marked with the month of manufacture and installation date. They should be replaced before 75 per cent of their shelf life has expired.

Power supplies and cables should be of correct voltage and sufficient capacity to supply the equipment they serve. The mains supply should be a key-switched or unswitched outlet with indicator.

System handover

The installation should be inspected as the work proceeds so that parts difficult to examine later can be checked. Connections should be examined, and the securing of wiring, junction boxes and sensors inspected. Measurements should be made of all detector circuit resistances and insulation, and current taken from each power supply, the readings being recorded in the record book.

On completion, the inspection should confirm that all parts conform to the relevant British Standard: that mains power supplies are permanently connected and correctly fused, batteries are dated and of adequate capacity; that an alarm does not result from automatic changeover between mains and battery operation, that anti-tamper circuits, and all detectors including perimeter switches, movement sensors, beam interruption devices, and inertia detectors are each properly functioning, that zone switching is operating correctly; that the exit/entrance circuit is working and the timing is correct. Exit time should be recorded. (*A check should also be made that adjacent sensors are not easily triggered on the exit route. Panic buttons should also be tested.*)

If self-activating sounders are used, their operation should be checked by removing the hold-off voltage. The system should be switched on and the system tested by operating a sensor. Check all sounders.

Telephone signalling equipment should not be connected to a system having movement sensors, until a minimum of 7 days false-alarm-free functioning have elapsed. This period should be before the official handover. The various functions and safeguards of the dialling apparatus as listed in a previous section should be tested, and the results recorded.

The tests and handover should be by a person other than the installation engineer. He should be sure that the owner is fully informed of the system's operation, has written instructions, and signs to that effect. Police should be informed that a system has been installed especially if it includes telephone signalling, and be advised by the owner of the names and addresses of keyholders.

Routine maintenance

A visual inspection for damaged sensors and cable, and loose or broken controls and indicators on the control panel, should be made. There should be a check on security of door and window fastenings. The tape and mechanism of telephone signalling equipment should be tested and cleaned if necessary, but precautions should be taken that a call is not inadvertently triggered. A test call can be made only by prior arrangement with the security control station. All batteries should be checked, cleaned and greased as required, specific gravity readings of wet battery electrolyte should be taken and corrected. (*Apart from this, the checks advised in the code for routine maintenance are similar to those made prior to handing over a new system.*)

False alarms

A four-week log should be kept by the alarm company in which all false alarms are entered. Troublesome systems can be identified and investigated by a specially appointed Alarm Systems Performance Executive (ASPE). This should reveal recurring defects in equipment, or in installations by particular engineers.

At each of four successive unexplained false alarms during a four week period, the responsibility is passed from engineer to supervisor, to nominated engineer, to manager, then to regional manager; the ASPE being brought in on the third and fourth occurrences.

A diagnostic check-list is given, which should be followed for each incident after the initial one. First, ensure that correct operation procedure is being followed by all persons using the system, and that no changes have been made to the structure, electricity supply, heating systems, automatic lighting control that could affect the system, or any source of r.f. interference.

After these, check the control equipment and power supply including batteries. Walk-test and check the sensitivity of all movement detectors. See that nothing can obstruct beam detectors (*including dirt on the active faces*). Check that protected doors and windows are firmly secured when closed. Measure loop resistance for steady readings that conform to those previously recorded. (*A check that pressure mats are clear of standing objects should also be made.*)

The operation and connections to telephone equipment and sounders should be checked. (*The latter may seem superfluous when investigating false alarms, but a poor connection in the hold-off voltage of a self-activated bell could cause an alarm condition.*)

Check anti-tamper circuit connections and the exit circuit including associated lock or key switches.

(*Not all of these steps would be necessary in a given situation. The latching indicators on the control panel and on movement detectors will eliminate many of the possibilities. So for example, there would be no need to check anti-tamper connections if the detection loop is latched.*) Commercial details and definitions, with responsibilities of an alarm company and the subscriber are also given in the code.

22 *Installing fire alarms and maintenance*

The type and location of the sensors should be decided according to the type of fire risk and area to be covered as shown in Chapter 16. To briefly recap, ionizing smoke detectors are the fastest acting where the combustion products do not generate dense smoke and in many cases even when they do, as lighter combustion products are often given off in the incipient stages of a fire. Optical smoke detectors can be used where heavy smoke is quickly produced.

Heat detectors are slower but must be used where dust, steam and fumes normally in the atmosphere, would give rise to false alarms. Of the two types of heat detectors, one responds when the temperature reaches a particular level, usually 135° F (57° C), while the other is triggered by a rapid temperature rise. The latter often incorporates the former as well so is usually the best type to use (Figure 91). It should certainly be used for cold areas where a fire may take some time to raise the temperature to the fixed point of 135° F.

The ability of a detector to sense smoke or heat generated some distance away, depends on the sideways dispersion produced when smoke or hot air reaches the ceiling. They spread out at ceiling level, so reaching detectors not

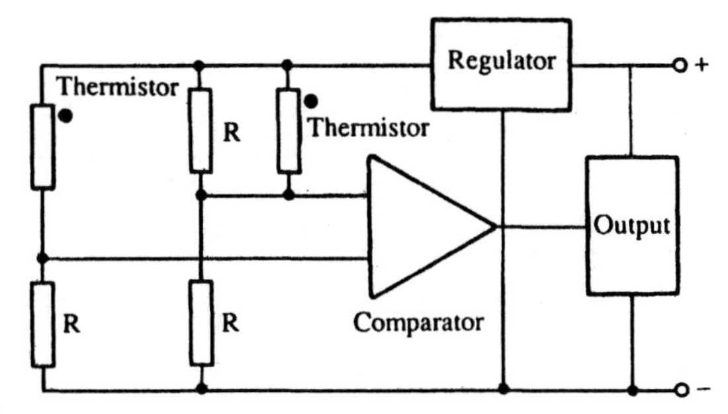

Figure 91 Circuit of a rate-of-rise heat detector. A fixed voltage derived from a regulator is applied to two chains containing thermistors. One responds to temperature change more slowly than the other so that an unequal voltage is applied to the comparator which triggers the output.

immediately above the fire. It follows that the detectors should be mounted on the actual ceiling and not on beams or girders below it.

As an obstruction, such as a beam, restricts the sideways spread, the path around the obstruction must be allowed for when measuring the distance between sensors (see Figure 70). Any obstruction deeper than a tenth of the floor to ceiling height should be regarded as a wall and the areas on either side as separate rooms. Distance from any point to the nearest smoke detector should not exceed 25 ft (7.5 m), or to the nearest heat detector, 17 ft (5.3 m).

With pitched roofs or north lights, the detectors must be fitted to the apex, or at least not more than 6 in (150 mm) below for a heat sensor or not more than 2 ft (600 mm) below for a smoke detector. Any pitch shallower than those measurements can be regarded as a level ceiling.

Bells should be sited so as to produce no less than 65 dB anywhere in the protected area, and at least 5 dB higher than any background noise likely to persist for longer than 30 s. Where there is sleeping accommodation, the sound level should be not less than 75 dB at each bedhead.

Wiring

The danger to wiring is not tampering, as it is with intruder alarm systems, but destruction from the fire itself. Wiring to the sensors is not vulnerable though, as the detector should have sensed the fire and triggered the alarm long before its wiring is affected. Once triggered, the alarm is maintained by the control panel so the sensor or its wiring plays no further part. Detector wiring can thus be any suitable cable which is polarity coded because the polarity of the sensors must be observed.

Most of the detector wiring will be at the same height as the detectors, that is, at or near ceiling level except for down-drops to manual call points. To protect against mechanical damage, down-drops or any wiring below 8 ft (2.5 m), should be sheathed or run in conduit.

If the open-circuit monitored system is used, the detectors are connected in a chain, each having a pair of input terminals for the incoming wires, and a pair of output terminals for the wires to the next detector. As the input and output pair are linked together it may be wondered why there are separate terminals. One reason is to give full monitoring.

If the input and output wires are twisted together on the same terminal, a loose terminal may not show up because the circuit could still be continuous to the next detector. The sensor having the loose terminal could thus be inoperative without giving any warning. With two pairs of terminals, any loose connection breaks the chain and thereby generates a warning signal at the panel. The two pairs of terminals should therefore always be used if fitted.

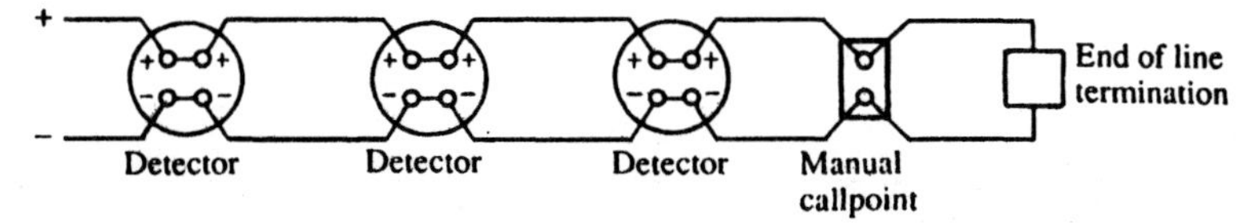

Figure 92 Two-wire 24 h monitored fire detection circuit. A small a.c. current flows through the end-of-line termination and a fault indication is triggered if it ceases. Polarized detectors go low resistance when triggered. Each takes a small current when quiescent so the number per zone is limited to the maximum rated current. Unlimited number of call points can be connected.

Monitoring is accomplished by superimposing an a.c. voltage on the detector line which passes current through a capacitor and diode shunt at the end of the chain. Any break in the chain interrupts the current and signals a fault. The end-of-chain components can be conveniently connected across the output terminals of the last sensor (Figure 92).

If follows that no spurs can be connected at any point along the chain as they would be unmonitored. If an end-of-circuit shunt was connected at the end of a spur as well as the main chain, the panel could not distinguish between them. A break on either the spur or that part of the main chain beyond the spur connection would not trigger a fault signal because the other shunt would maintain the closed circuit. For the same reason, one chain should not be connected to the start of the other, in a parallel configuration, the wiring must always be in the form of a single continuous chain to each zone.

To avoid shunting the a.c. monitoring current by excessive cable capacitance, or impeding either monitoring or alarm currents by high cable resistance, some control panel makers specify maximum values for these, but it is unlikely that they would be exceeded in a practical installation. For example, one panel specifies a maximum resistance of 150 Ω. Now 16/.02 mm flat polyvinyl chloride (pvc) cable has a resistance for both conductors of 7.2 Ω per 100 m. A length of 7000 ft (2100 m) would therefore be required to exceed 150 Ω. Even a thinner cable, 7/.02 mm has a resistance of 16.4 Ω, allowing up to 3000 ft (900 m).

Cable capacitance is even less likely to be exceeded. The same specification stipulates a maximum of 0.5 μF (500,000 pF). The 16/.02 cable has an approximate capacitance of 27 pF/ft (90 pF/m). This would require 18,500 ft (5500 m) to exceed 0.5 μF. Such lengths would be very unlikely on a single zone. Thinner cable such as 7/.02 has even lower capacitance.

The number of sensors must not exceed the maximum current that the control panel can supply for each detection zone. A nominal 0.1 mA can be reckoned for each sensor, so a 4 mA zone rating means that forty detectors can be wired per zone.

Actually, the current taken by most sensors is rather less than 0.1 mA. So if a few more are needed than the maximum number usable at the 0.1 mA rating, it may be possible to accommodate them. The exact current taken is usually published by the sensor manufacturer, so this should be consulted in cases of doubt and the number calculated accordingly. There is no limit to the number of manual call points that can be wired to any zone in addition to the sensors as these do not pass any current. Call points and sensors of all types can be wired in any order, providing they are designed for the type of circuit in use. Normally-closed sensors and call points can only be connected in a loop.

Another point to watch is that some sensors, particularly those intended for inclusion with domestic intruder alarms may be rated at 12 V although the standard voltage for fire alarm systems is 24 V. The voltage of all sensors should be checked and must correspond to that specified for the panel.

Bells should continue to sound for as long as possible and not be inactivated at an early stage by the fire. Wiring to them should therefore be heat resistant to a reasonable degree. Silicone-rubber insulation can withstand temperatures up to 390° F (200° C) but is rather expensive. PVC electrical installation cables or elastomer-insulated textile braided wiring can be used providing they are protected in conduit. Total cable resistance should be low enough to give negligible voltage drop to the furthest bell at full load.

Power cables to the control panel are also vital and should be likewise protected. Just as with the intruder alarm system, connection to the mains supply should be by a non-switchable connector and never by means of a plug and socket.

Testing

When the installation is completed all parts should be tested. This is a rather time consuming task with a large installation, but essential. If the sensors are easily removable from their wired bases, they should be, and each detector zone meggered for leaks. This should not be done with any detector in circuit. Next, the standby batteries should be connected. A fault indication on the panel should show that there is no mains supply. The mains are now connected and the output of the panel switched from the alarm bells to the test buzzer.

Each detector must now be tested in turn. Smoke detectors can be checked by holding a piece of smouldering sash cord near the vents and gently blowing smoke into them. Chemically generated 'smoke' from aerosol cans can be corrosive and could coat the inside of the detector chamber, so it is not recommended. The smouldering cord gives the minimum of undesirable side effects including soiling of ceiling decor.

For heat detectors, a portable hair dryer is the most practical method of testing. Some dryers just about get to 135° F (57° C) which is the operating point for fixed heat sensors. So, check the dryer first with an accurate thermometer to see that it does attain this temperature. It may take several minutes at the highest setting to get there. Do not overlook the possibility that a dryer that easily made it in mild weather, may not do so when the weather is cold.

In the case of the rate-of-rise heat detector, the rapid increase from the ambient temperature when a hot air dryer is directed into it should trigger it quite quickly. This incidentally demonstrates the faster response of this type of sensor.

Manual call points can be tested by various means depending on the model. Some can be opened from the front thereby releasing the pressure on the button, while others have a special key which moves the glass relative to the button. Whenever a sensor or call point is actuated, the control panel latches and so it must be reset before the next sensor is tested. A two-man team with radio communication will considerably expedite the job, especially in a large building where the control panel may be some distance from the furthest detectors.

After all the detectors and call points have been tested, the furthest detector in each zone should then be removed. This tests the monitoring circuit which should show a fault for that zone. Finally, after switching on the main alarm bells, a call point can be actuated and each bell checked to see if it is working. A sound level meter should be used to check that the sound level everywhere is at least 65 dB and above 5 dB greater than the ambient noise. Where there is sleeping accommodation, the reading should be 75 dB at each bedhead with the intervening doors closed.

A record book should be provided similar to that used for intruder alarm systems, in which technical details can be recorded, and space made available for entering service and maintenance events.

Maintenance

Daily

The effective operation of a fire alarm system on which lives could depend, should be regularly checked and tested. As the system is continually and automatically monitored for faults, a daily examination should be made on the panel to check the status of its fault indicators. Any fault should be recorded in the log book and reported to the management.

Weekly

Monitoring checks the various circuits but not the actual operation of the detectors, sounders and panel. Once a week, a detector or call point should be triggered and the sounders checked; a different detector being chosen each week until all have been tested in turn. The standby battery terminals should be examined for corrosion and cleaned and greased if required. The date, time, and identity of the detector tested should be entered in the log. Faults should be noted and reported.

Quarterly

Quarterly tests should be made which, in addition to those carried out weekly, simulate faults by disconnecting the mains supply, and by removing the final sensor on each zone. Correct operation of the fault indicators is thereby checked. The quarterly system tests should be carried out without the mains supply, and the battery voltage should be checked with all sounders operating. Like humans, batteries deteriorate quickly when idle and are preserved by at least occasional use. The quarterly test should help provide the activity they need together with the subsequent recharging by the panel. BS 5839 suggests that a test certificate be made out for the quarterly test.

Annually

Every year the current quarterly test should be augmented by cleaning every smoke detector and testing each one. A visual examination of all fittings and accessible wiring should also be made.

Cleaning detectors

For cleaning ionizing smoke detectors, three brushes are recommended, a stiff brush and two soft nylon brushes. One nylon brush should be designated the 'wet' brush and used to apply methylated spirit, and the other kept dry.

The stiff brush is used to clean dust and dirt from the exterior casing and anti-insect mesh. Any grease will need to be removed with spirit and the wet brush.

On dismantling, the chamber can be cleaned first with the wet brush and then deposits removed with the dry one. Care must be taken in cleaning the radioactive element to avoid excessive radiation, although the amount present is minute. Precautions are: holding brushes near their ends so that the fingers do not get closer than 2 in (50 mm) to the source, not letting the eyes get closer than 6 in (150 mm) to it, and not inhaling any vapour from contaminated spirit. Use the minimum of spirit on the brush and throw away any left-over spirit into which the brush has been dipped. Clean the brush afterwards with fresh spirit. Clean the 'dry' brush with soapy water and allow to dry completely before reuse.

The exterior and the mesh of optical detectors are cleaned in the same way, but the chamber interior, lenses and optical units can be cleaned with a 'puffer' brush such as used to clean camera lenses. Especially watch out for light coloured dust on the interior walls of the chamber which could scatter the infra-red light and generate false alarms. Do not polish or buff up the walls which must be matt black.

As may be observed with room ionizers, dirty deposits soon develop due to the attraction of dirt particles by ions. So ionizing detectors get dirtier and need wet cleaning, while optical ones usually need only an internal dust out. Heat detectors need little if any cleaning at all, but could be given a dust over and a check to see that dirt or grease has not built up over a long period.

Detectors should never be painted, as this would modify their thermal characteristics and their response to heat.

23 Reference information

Standby batteries

These can be either the primary (non-rechargeable) or secondary (rechargeable) types. Primary batteries for security systems can be of the zinc-carbon type which is the same as ordinary torch cells, or alkaline.

A major factor with the zinc-carbon type is the polarizing effect. Hydrogen is generated by the chemical action of the ammonium chloride electrolyte on the zinc case, which collects around the positive carbon electrode thereby insulating it and blocking any further activity. The cell is then said to be polarized and needs to be rested to allow the manganese dioxide depolarizer to absorb the hydrogen. Polarizing occurs quickly during heavy discharges, but does not occur at all when the discharge is small and the depolarizer can absorb the hydrogen as it is produced. High output types of cell have a larger quantity of purer manganese dioxide to take up the hydrogen more rapidly.

Alkaline cells use compressed manganese dioxide instead of carbon for the positive electrode and potassium hydroxide for the electrolyte. This enables them to deliver heavier currents for longer periods before polarizing. They have lower internal resistance and a capacity some five times that of the zinc-carbon type. A major factor with security systems is the much longer shelf life.

Suitable primary cells are quite adequate for domestic or small business alarm systems, but the continuous drain of the anti-tamper loop and the loop current taken by several zones make the secondary cell preferable for the medium to large installation. This is particularly so when passive infrared detectors (PIRs) are part of the system as these take a continuous current of around 15 mA.

Secondary batteries can be of the lead-acid type as used in motor vehicles, or the nickel-cadmium (NiCad) type. Their capacity is rated in ampere hours, which is the number of hours that a given current in amps can be sustained. Thus a 10 A/h battery would give 1 A for 10 h or 2 A for 5 h, but the capacity falls with higher discharges so that a 4 A current for the above battery would probably last for less than 2 h thus giving a capacity of under 8 A/h. Batteries for control panels are usually from 1.2 to 6 A/h, note that some panels specify a maximum capacity. A trickle charger is usually built in to the panel so that the batteries are always kept fully charged.

The lead-acid type consists of positive plates of lead peroxide interleaved with negative plates of spongy lead immersed in a dilute solution of sulphuric acid. They have high capacities and can deliver high currents for sustained periods. The main disadvantages are weight, size and the danger of acid spillage. Another problem is that active material from the plates flakes off with age and eventually forms a conductive bridge across the bottom of the cell. Some models have a sump at the bottom to collect the debris and so delay the build-up, but this reduces the capacity/volume ratio.

Acid spillage is overcome by using jelly-acid, or porous separators that absorb most of the electrolyte and also sealing of the container. The cell voltage is 2 V which drops fairly linearly with discharge time. When trickle-charged the voltage is maintained at the full 2 V, but if used to power remote signalling equipment, the battery must be recharged when cells fall below 1.8 V. Most control-panel standby batteries are of the lead-acid type. Panel charging voltage is 13.6 V.

The NiCad (more accurately NiCd) cell comprises a strip of perforated sintered nickel mesh and one of sintered cadmium rolled up together and separated by a strip of insulating material impregnated with a solution of potassium hydroxide. The voltage is 1.25 V per cell, so ten cells are required to obtain 12 V as against six of the lead-acid cells. They are thus more expensive.

Most types of cell deteriorate if inactive, an important factor in alarm systems. The NiCad does not, although the charge will leak away with time. Charge retention is dependent on temperature, at 68° F (20° C) the loss of charge is 10 per cent in 10 days. At freezing point the loss of 10 per cent is over 30 days, but at high temperatures such as 86°F (30° C) some 20 per cent is lost in 10 days. Voltage is maintained over most of the discharge period, so even when some of the charge has leaked away, the voltage remains the same and equipment is at full efficiency.

Unlike most other types, the cells perform down to –8° F (–22° C). At freezing point the capacity drops by only 20 per cent. This is important in the operation of remote signalling equipment and outside bells.

The capacity of a NiCad cell is also governed by the charging rate, the lower the current, the higher the capacity. The rate is denoted by the letter C combined with multipliers or submultipliers. Thus C/1 means the rate at which a cell will be charged in 1 h and is the basis from which others are calculated. A charging rate of C/4 is a quarter that of the 1 h rate but the time required to charge is 5 h and so the capacity is $5/4 = 1\frac{1}{4}$ times that of C/1. The C/4 or 5 h rate is the normally specified capacity.

Very high rates are given by the multipliers 2C, 4C and so on, these being twice and four times that of C/1 with the capacities being less still. For example, a NiCad D-cell has a 4 A/h capacity at C/1 which means it will take 4 A for 1 h to charge it. At C/2, which is a charge of 2 A it will charge for 2¼h giving a capacity of 4½ A/h. However, at 2C which is a charge of 8A, the time is 27min giving a capacity of 3.6 A/h.

An important feature of NiCads is that overcharging is permissible only at low charging rates. Thus trickle-charging can be allowed only at C/8 or less. To fully charge from the discharged state takes 12 hours at C/8.

The same applies to discharging, with greater capacities in ampere-hours being available at lower discharge rates. The small discharges required by alarm systems thus result in maximum capacities being achieved.

Table 4 Table of charging times of different C rates and maximum charges. Over 4C the maximum is the same as the nominal and careful timing is required to avoid damage by overcharging.

Rate	*Charge time*	
	Nominal	*Maximum*
C/40 (minimum)	70.0 h	Indefinite
C/10	14.0 h	Indefinite
C/10 (mass-plate)	14.0 h	42.0 h
C/8	12.0 h	Indefinite
C/4	5.0 h	6.0 h
C/2	2.25 h	2.5 h
C/1	1.0 h	1.25 h
2C	27.0 min	30.0 min
4C	12.0 min	12.0 min
8C	5.0 min	5.0 min

There is also a minimum charge rate, which for sintered cells is C/40. This would take 70 hours for a complete charge from fully discharged. A rate less than this does not charge the cell at all. This must be considered when selecting the size of a NiCad that is to be trickle-charged.

For example a battery consisting of D cells having a C/1 of 4A, has a C/40 rating of 100mA. So the trickle charge must be greater than 100mA or the battery will not be charged. If the charging current supplied by the equipment is less, smaller cells must be used. The C-cell has half the capacity of the D-cell and so can be kept charged at 50mA, which is its C/40 rating.

Not all NiCads are of the sintered type, some are *mass-plate cells*. These include button cells and those designed for mounting directly on to printed-circuit boards. The maximum rate for these is C/10 which takes 14 hours to fully charge, but for these, the maximum time at C/10 is 42 hours. Continuous trickle-charging should be at no less than C/100.

If a NiCad is charged from an ordinary charging circuit, the current decreases as the cell voltage rises, so the rate of charge varies and it is not possible to accurately calculate the charging time. Also because of its low internal resistance, a very heavy current can flow at the start of a charge when the cell is fully discharged. To avoid these effects, the charging circuit should be of the constant-current type which maintains a steady current irrespective of cell voltage. This does not apply to trickle-charging where there is no danger of overcharging and the cell is never fully discharged. The NiCad cell exhibits a 'memory effect' whereby the capacity is permanently reduced if not fully discharged before recharging. The new Nickel–Metal–Hydride cells overcome this, and offer a higher capacity for the same size.

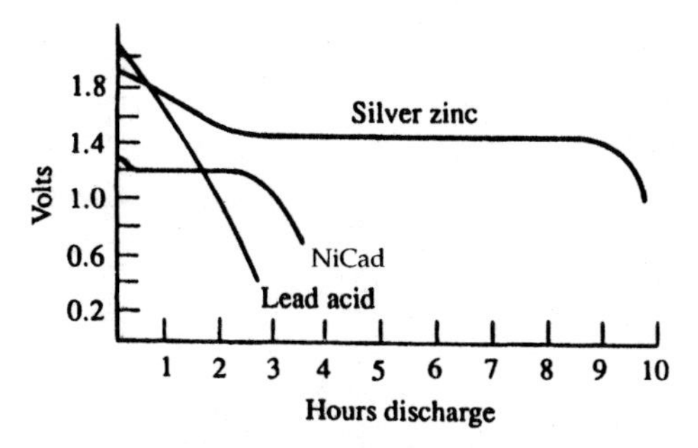

Figure 93 Discharge voltages of comparable different type secondary cells.

Another type of cell is the silver-zinc, which has a negative electrode made of zinc, and a positive of silver-oxide. The electrolyte is potassium hydroxide, the same as with NiCads. The weight and volume for a given capacity is the lowest for any secondary cell in general use, being about a third of that of the lead-acid type. Furthermore very high currents can be

Table 5 Average charging cycles for different cell types

Cell type	*Number of cycles*
Lead-acid	500
NiCad	2000
Silver-zinc	150

taken from the cell without damage, and they can be recharged very quickly.

Voltage is virtually constant at 1.5 V over the whole discharge period; this is useful, but can be matched by the NiCad although it has a much shorter discharge time. Silver-zinc cells are costly, but their main snag is the limited number of charge/discharge cycles. These are compared in Table 5.

Standby battery circuits

Most equipment has a built-in arrangement for switching over to a standby battery if the mains supply should fail. Should this be lacking it can easily be provided by using either a relay or a couple of diodes.

Figure 94 shows the relay circuit. The d.c. supply energizes the relay which closes the contacts that connect the supply to the load. If the supply fails, the relay releases the contacts thereby switching the battery to the load. The supply is maintained by C1 during the changeover, otherwise the momentary break could trigger a sensitive alarm circuit. D1 prevents C1 discharging through the relay coil which would keep it energized and thereby delay the switching until after C1 is exhausted.

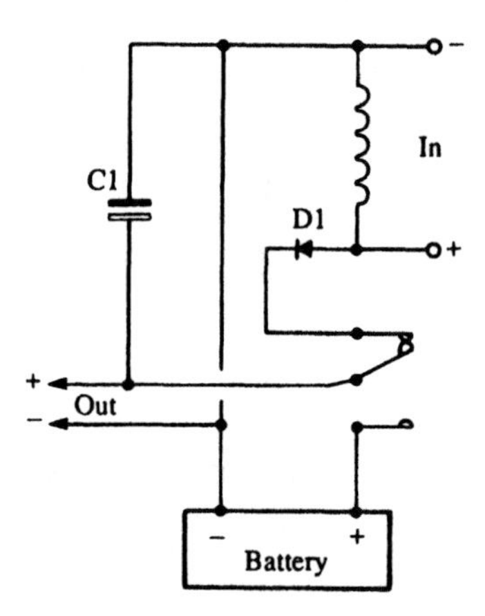

Figure 94 Relay-operated changeover to standby battery circuit.

A simple but effective changeover circuit is shown in Figure 95. The supply from the power circuit must be about 1 V higher than that of the battery. When operating from the mains, current flows through D1 to the load. It is prevented from flowing into the battery by D2. As the supply voltage is higher than that of the battery, D2 is reversed biased and so does not pass current from the battery to the load.

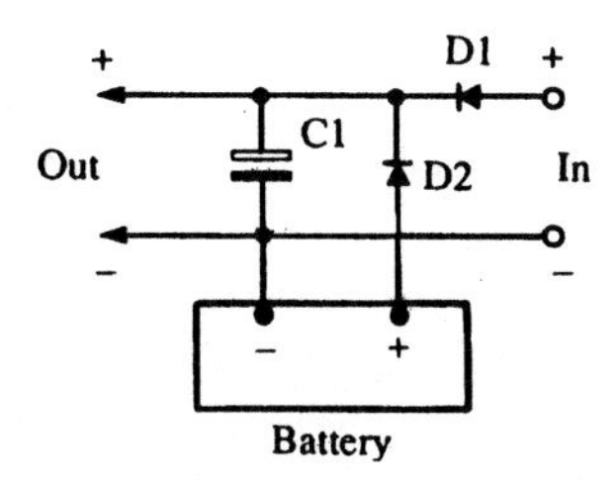

Figure 95 Diode-operated changeover circuit.

If the supply voltage disappears, D2 is no longer reverse biased and current flows from the battery to the load. None flows back into the supply circuits because of the presence of D1. The diodes thus isolate the battery and supply from each other and provide a near instantaneous changeover. The capacitor C1 smooths the change to the lower battery voltage.

The diodes produce a voltage drop of about 0.6 V irrespective of load current, so this should be allowed for. In the relay circuit there is a drop in the mains-derived voltage but not the battery volts as there is no diode in series with the battery. With the second circuit there is a drop of both battery and mains-derived voltage.

Both these circuits are intended for use with primary batteries that need to be isolated from the supply. For secondary batteries for which trickle-charging is required, Figure 96 can be used. The supply current travels through D1 to the load, while charging current for the battery passes through R1, which is chosen to limit the current to the required charging rate. D2 prevents the battery discharging through the supply circuit.

D3 is reverse-biased by the higher supply voltage and so prevents the battery discharging to the load when the supply is present. If the supply fails, D3 becomes conductive and passes current from the battery, but it is prevented from flowing back to the supply circuit by D1.

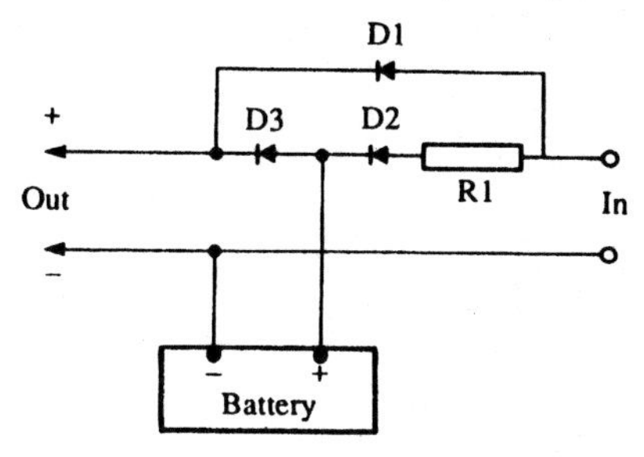

Figure 96 Changeover circuit that also trickle charges secondary battery.

Voltage regulators

Some active detection devices such as passive infra-red detectors (PIRs) may trigger if there is a sudden change in the supply voltage. While a large-value capacitor connected across the d.c. supply line will smooth the change and make it gradual, a regulated circuit is better and will keep the voltage to within 0.5 per cent at 12 V.

A popular regulator i.c. which has built-in overload, thermal and short-circuit protection is the 78, 79 and 78S series. The 78 is rated at 1 A, positive output, the 79 at 1 A negative output, and the 78S, 2 A positive output. There are many other types but for most alarm system equipment, the 78 is the most suitable. Suffixes denote the output voltage.

The devices are encapsulated with a three-pin lead out which is the same for the 78 and 78S series, but not the 79. A heat sink is required which in most cases can be achieved by bolting the device to a main member of a metal case. A little silicone grease smeared on the back of the mounting lug before it is fitted will improve the heat-transfer.

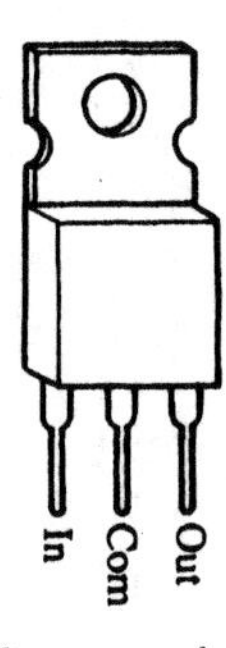

Figure 97 Pin connections of type 78 range of voltage regulators.

The pin configuration is shown in Figure 97. The metal fixing lug is connected to the common earthed pin with the 78 and 78S regulators and so for most circuits can be bolted directly to an earthed metal chassis or case without insulating it. A regulator circuit is shown in Figure 98.

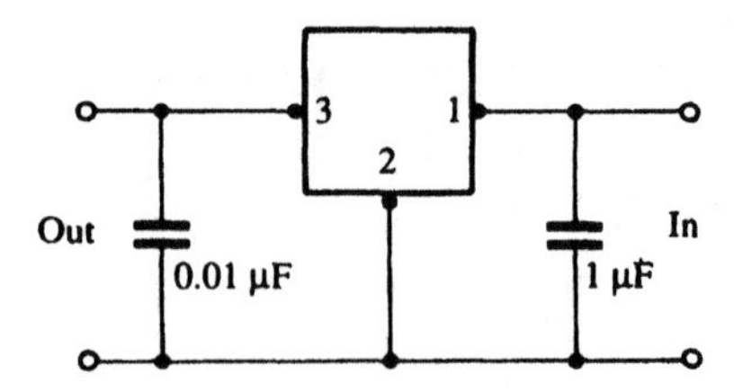

Figure 98 Regulator circuit using type 78 regulator.

Table 6 Regulator data

Device	*Output voltage (V)*	*Input Voltage (V)*	*Equivalents*
7805	+5 (±0.2)	7.0–25	μA7805UC; LM340T-05; MC7805CP; SN72905.
7812	+12 (±0.5)	14.5–30	μA7812UC; LM340T-12; MC7812CP; SN72912.
7815	+15 (±0.6)	17.5–30	μA7815UC; LM340T-15; MC7815CP; SN72915.
7824	+24 (±1.0)	27.0–38	
7905	−5 (±0.2)	7.0–25	
7912	−12 (±0.5)	14.5–30	
7915	−15 (±0.6)	17.5–30	μA7915UC: LM320T-15 MC7915CP.
7924	−24 (±1.0)	27.0–38	
78S05	+5 (±0.2)	8.0–35	
78S12	+12 (±0.5)	15.0–35	
78S15	+15 (±0.6)	18.0–35	
78S24	+24 (±1.0)	27.0–40	

Cable resistance

The resistance of long runs of cable can be significant, especially for bells or sounders on 12 V intruder alarm systems as first, the current taken by these and the consequent voltage drop, is greater than with 24 V or 48 V circuits. Second, that drop constitutes a higher proportion of the original voltage with a low voltage system.

Most bells have a specified current rating, but this can be misleading because it is an average value that may be indicated on a well damped analogue meter. When the contacts are open the current is zero, but when they close, it is much higher than the rated average. Now, it is the voltage across the coil and the current flowing when the contacts close, that are the important values, because they produce the power that impels the striker. Therefore, it is the current that matters when considering the effect of series cable resistance, not the specified average.

This current can be approximately calculated from the coil resistance, although it will be somewhat less than this due to the coil inductance. For example, one bell having an average rating of 80 mA, has a coil resistance of 48 Ω. The closed-contact current at 12 V is thus 250 mA, but allowing for inductance, possibly around 200 mA.

Table 7 sets out the various wire sizes with the total resistance per 330 ft (100 m) of a twin cable. (A single wire is therefore half the resistance shown). Also given for quick reference, is the voltage drop over 100 m for a current of 200 mA. Note that some low voltage bells take more than this.

Table 7

Wire size	*Resistance/100 m (Ω)*	*Voltage drop at 200 mA (V)*
1/0.1	438	87.6
1/0.2	115	23.0
1/0.6	12.8	2.56
1/0.8	7.2	1.44
7/0.2	16.4	3.28
13/0.2	8.8	1.76
16/0.2	7.2	1.48
24/0.2	4.8	0.96
32/0.2	3.6	0.72
30/0.25	2.4	0.48
40/0.2	2.9	0.58
0.5 mm^2	7.2	1.48
1.0 mm^2	3.6	0.72
1.5 mm^2	2.4	0.48
2.5 mm^2	1.4	0.28

Video cable impedance

All cables possess four parameters which are: resistance and inductance in series and capacitance and conductance in parallel. The latter arises from leakage across the insulation which is never perfect. It is the reciprocal of resistance and is expressed in mhos or siemens. A leakage of 1 megohm gives a conductance of 1 micromho (or 1 microsiemen).

A video cable thus has a characteristic impedance which depends on these parameters. The formula is:

$$Z = \sqrt{\frac{R + 2\pi fL}{G + 2\pi fC}}$$

in which R is the resistance in ohms, L is the inductance in henries, G is the conductance in mhos, C is the capacitance in Farads and f is the frequency.

If L/R is equal to C/G the frequency element in the formula cancels and the impedance becomes independent of frequency. The impedance, unlike resistance does not depend on cable length because all parameters are of the same proportion to each other, whatever the length.

Cable impedance should match that of the source, which is the camera, and should be terminated at the end by a resistor of the same value, to avoid reflections back along the cable. The standard is 75 Ω, but there is also 50 Ω cable. The impedance of an unknown cable can be estimated by measuring its capacitance; 75 Ω cable has a capacitance of about 60–68 pF per metre, whereas the capacitance of 50 Ω cable is generally between 85–100 pF per metre.

Video pick-up devices

The standard video camera pick-up device has for many years been the vidicon. This is the generic name given to a group of similar tubes differing mainly in the material used for the target, each being known by a particular trade name. The target is a photoconductive layer, antimony trisulphide in standard vidicons, which is deposited on the end of the tube like the screen in the cathode-ray viewing tube. A cathode heated by a filament, generates a stream of electrons which is deflected over the target area by a scanning magnetic field set up by coils fitted around the tube.

The target is connected to a high positive voltage which attracts and returns the negative electron beam to the supply circuit. Surface resistance of the target varies according to the light falling upon it, so the beam current changes in value as the beam scans across the picture image which is focused on the target. This produces voltage variations over a load impedance, which after insertion of sync pulses to identify the end of each scanning line and frame, constitute the video signal.

Sensitivity of the device can be changed by altering the target voltage. Too great a sensitivity in bright light results in over-contrast, often called 'soot-and-whitewash', so the voltage and sensitivity can be set by means of a simple potentiometer. Together with different lens apertures, a wide range of light conditions can be accommodated.

Illumination levels are measured in lux, or sometimes in foot-candles; the relation between these is 1 foot-candle = 10 lx. Around 10–20 lx is the minimum scene illumination to give a viewable picture with a standard monochrome vidicon of 2/3 in diameter. Other types using different target materials are more sensitive. The RCA *Ultricon* for example produces pictures at less than 1 lx. Bright moonlight gives about 0.3 lx, so the sensitivity can be judged from that. The variable sensitivity facility, however, is not available with these tubes.

For colour cameras, filters in the three primary colours are commonly used in front of the tubes, and these considerably reduce the amount of light falling on the target. Sensitivity for colour is therefore lower than with a monochrome camera. An illumination level several times greater is required to produce satisfactory pictures.

A major drawback with the vidicon is its vulnerability to burn-in. If a bright stationary scene is left on the target for a period, it will produce a permanent image which appears as a ghost with all subsequent pictures. Accidental exposure to a very bright light such as the sun for only a few seconds, will destroy that part of the target coating affected by it. Another snag is the long-lag characteristic, whereby images that move or are panned at low-light, high-sensitivity settings, leave trails behind them.

The Ultricon has a much shorter lag-time but does not have such a good definition as the standard vidicon. Toshiba's *Chalnicon* tube which has a cadmium selenide target is not quite as sensitive as the Ultricon nor has it so short a lag-time, but it has a better resolution. It is also better than the standard vidicon for lag-time and sensitivity.

Another version of the vidicon is the *Newvicon* made by Matsushita which has a target made of zinc telluride and cadmium telluride. Its lag-time and sensitivity is better than the standard vidicon and Chalnicon but not as good as the Ultricon; resolution though is better than the Ultricon. However it is more prone to burn-in than the others, with the exception of the standard tube.

Spectral response is different for all these tubes, and the prices are several times that of the standard vidicon. So there is no clear winner out of these, and camera designers choose the features that they think best, possibly making up for deficient characteristics in the camera circuitry.

The CCD

There are two problems that all of these tubes have in common. One is size; although not bulky, they do take up room when scanning coils, yokes, and ancillary items are included, especially with colour cameras where there is more than one tube, plus filters. The other is current consumption.

These are not major problems as far as security cameras are concerned, but they can be for domestic video units especially those incorporated in camcorders. To avoid them, the charge coupled device (CCD) has been developed, but it has other advantages which make it useful for security work too.

The CCD consists either of a single line or a lattice of cells consisting of closely spaced metal-oxide semiconductor (MOS) capacitors. These are formed on a silicon substrate on which is deposited a thin layer of silicon dioxide. On this is mounted the individual aluminium elements.

These are arranged in a linear group of three, and a pulse voltage with a trailing edge is applied across the silicon substrate and aluminium element of each cell in turn. This creates a tiny area depleted of electrons, like a well, under the pulsed cell. However, light falling on the cell releases electrons which partly fill the well in proportion to the amount of light.

The next pulse on the adjacent cell creates a well there, while the trailing edge of the first pulse reduces the depth of the first one. The electrons in the first well thereby spill over into the second. This in turn decreases in depth as the trailing edge of the second pulse passes, so tipping the charge into the well under the third cell, which has just been formed by the arrival of the third pulse.

By now the first pulse is completely clear and the well has disappeared, so all is ready for the next trio of pulses. The third cell discharges its contents into the first cell of the next group, and so on down the line. Thus at the end of each line there appears a consecutive string of charges from each group ready to be taken off to form one line of video signal.

This is known as the three-phase CCD, other configurations using two or four cell groups are possible. Linear charge-coupled devices consist of just a single row of cells, so these must be optically scanned by moving either the image or the CCD across the other for each field. Although simplifying the CCD, it complicates the optical system. The more usual type is the lattice device which forms a rectangle of cells on which the whole of the image is focused.

The resolution depends on the number of picture elements or pixels, which in this case is the number of capacitors on the substrate. For example, one CCD, has a matrix of 512 × 320, which is 163,840 cells.

It can be appreciated that with numbers of that magnitude it would be virtually impossible to produce a CCD with every single cell perfectly operative. Faulty cells can produce a white or black spot, or more seriously, if they refuse to pass on the charges received, can result in a blank line. The blemish specification therefore, is an important part of the CCD parameters.

As for sensitivity, the CCD device is better than conventional vidicons, although not quite as good as the Newvicon and Ultricon. While low current consumption and lack of bulk are advantages for portable video cameras, the major advantage of the CCD for security work is the freedom from burn-in. A continuously viewed bright scene will not become a permanent ghost image and the camera can actually be pointed at the sun without damage. It is also claimed to be free from lag.

Multiple cameras and sync

Whatever the pick-up device, the surface is scanned in horizontal lines of which there are 625 per frame in the European Standard, and there are 25 complete frames per second. Actually, the lines are not scanned consecutively, but first, all the odd-numbered lines are scanned, then the even ones are scanned in between them. Thus there are two vertical scans for each picture, giving 50 scans or fields per second. This *interlacing*, as it is called, gives reduced flicker on the reproduced picture.

The beam is deflected horizontally along the line by an oscillator circuit in the camera that produces a saw-tooth waveform, that is, one in which the voltage rises linearly then drops suddenly to the starting point. At the end of each line scan, a pulse is generated and added to the picture information. This synchronizes a similar oscillator in the monitor called a time-base, so that both run in step, and the scanning lines in the camera and monitor start and finish at the same time.

At the end of each field, another longer pulse is inserted so that the vertical scan time-base in the monitor keeps time with the one in the camera in the same way as does the line oscillator.

The exact speed or frequency of both line and field oscillators can be varied by pre-set controls in the camera. Pre-sets in the monitor, termed the line and frame hold controls, also need to be adjusted but minor errors in frequency are unimportant because the sync pulses from the camera pull the monitor oscillators into lock. Without these pulses, the picture would drift up or down and break up horizontally, even if set up precisely to start with, because all oscillators drift with temperature rise and other factors, if they are not controlled.

When more than one camera is to be used, as it often is with surveillance systems, the oscillators in each must be carefully adjusted so that their scanning frequencies are as near as possible to each other. While there will be minor differences, these will be within the synchronizing range of the monitor's time-bases. A stable picture will thus be produced when the monitor is switched between cameras without having to keep readjusting its hold controls.

If pictures from two different cameras are to be displayed on the same monitor screen at the same time to give superimpose, split-screen or inset effects, the camera oscillators must be identical in frequency and phase. This can only be accomplished by feeding sync pulses from the one into the other, or others if there are more than two. Thus one becomes the master camera that controls the oscillators of all the others as well as those of the monitors.

To enable this to be done, some cameras have *sync-in* and *sync-out* sockets. An alternative method which is used in TV studios is to have a separate sync generator which controls all the cameras. However, these effects are rarely required for security work.

Ultrasonic detectors

These work by utilizing the Doppler effect whereby the pitch of a sound changes if there is relative movement between the source and the listener. In this case, the movement is supplied by any moving reflecting surface in the area of detection, and the ultrasonic sound frequencies range from 23 to 40 kHz. To avoid frequency changes in the generated sound which could be interpreted by the detector as a Doppler shift arising from a movement, the oscillator must be very stable in frequency, and hence is usually crystal controlled.

The natural frequency of a crystal is dependent on its dimensions, and crystals of practical size cannot be made for the ultrasonic range. So the oscillator runs at radio frequency with crystal control, the output being divided down electronically to the required ultrasonic frequency.

The high-frequency tone is beamed into the protected area from a small transducer and received back by a microphone, the output from which is first passed through a high-pass filter to remove any audible sound. Any movement within the area produces reflections of a slightly different frequency from that of the original. The original is also present at the receiver, either from direct sound pick-up or by being mixed in electrically from the transmitter.

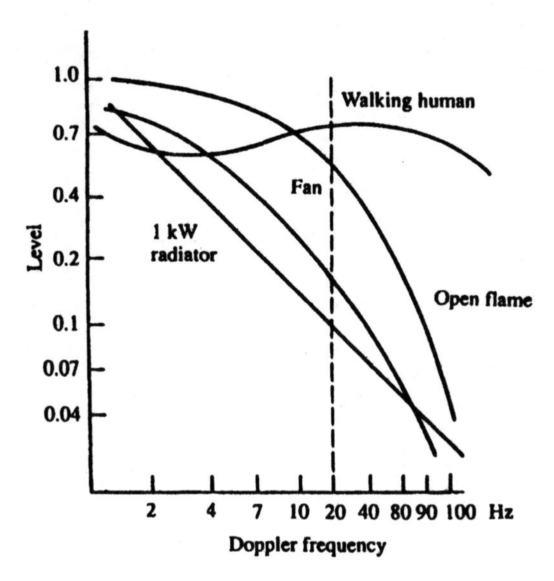

Figure 99 Doppler frequencies produced by ultrasonic detector for different moving objects. A discrimination against frequencies below 20 Hz, reduces false alarms due to non-human sources.

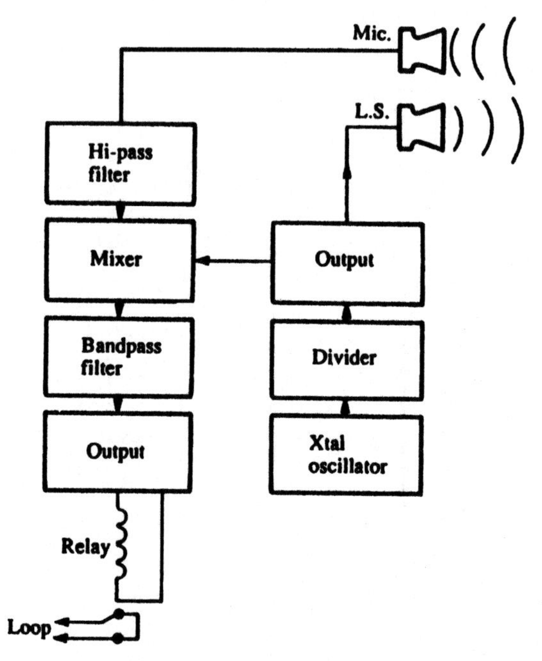

Figure 100 Ultrasonic detector block diagram. The output from a crystal oscillator is divided down and fed to a loudspeaker via the output stage. Reflected sound is picked up by the microphone, filtered to remove all non-ultrasonic sound, is mixed with the original and the difference signal, if any, is filtered to remove ultrasonic components and those below 20 Hz. The output operates a relay.

Two beat notes result, one the sum and the other the difference between the two frequencies. The sum is a very high frequency and is ignored, but the difference which is a comparatively low frequency, is passed through a low-pass filter to remove the original and reflected tones.

With a simple ultrasonic detector, the residual low-frequency difference signal, is amplified and made to operate the output relay. The more complex models include an analyser to discriminate between the Doppler frequencies caused by human movement and those caused by various other disturbances: Figure 99 shows these differences. Some immunity to false alarms is thereby achieved, but they can still arise from the high-frequency tones caused by vehicle brakes, gas flames, leaking air-lines, steam and door bells among other sources. Figure 100 shows the theoretical block diagram of the detector.

The range is up to 30 ft (9 m), and the coverage is in the form of a long narrow lobe in free air (Figure 101). This is considerably modified by reflections in an enclosed space. An important point, often overlooked, is

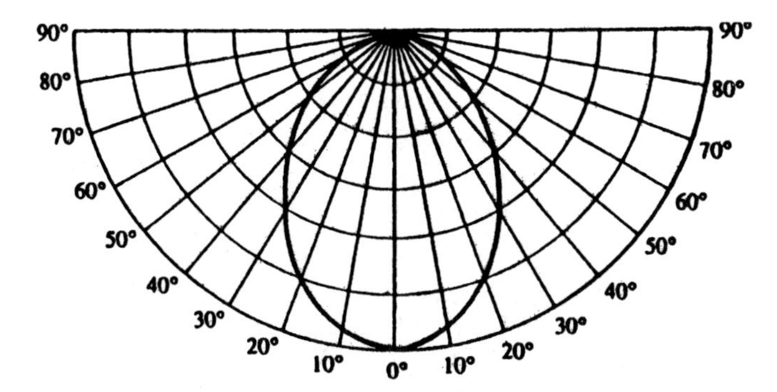

Figure 101 Polar diagram of the response of an ultrasonic detector.

that the unit must be fitted to an absolutely rigid support; pelmets, wooden partitioning and wall boards are unsuitable. Even low-level vibrations will cause the receiver to 'see' apparent movement of the whole area.

Two ultrasonic detectors should not be used in the same environment unless they are the same model and the makers stipulate that they can be so used. The different frequencies used in different models would produce beats and false alarms.

Microwave detectors

These also use the Doppler effect and work on the same principle as the ultrasonic detector except that very-high-frequency radio waves of the order of 1.5–10.7 GHz (1 GHz = 10^9 Hz) are used. The generator is known as a Gunn diode. Power is very low, typically 10 mW, so there is no danger of anyone being 'cooked' as with a microwave oven. (Could this be the ultimate intruder deterrent?)

The range is greater than that of the ultrasonic device, being up to 150 ft (45 m), and the beam will penetrate most solid objects except metal. This eliminates blind spots in warehouses and factories, but as penetration of boundary walls is possible, the system is liable to false alarms due to movements outside.

The range can be reduced by means of a sensitivity control, and when so limited, the shape of the beam is modified. The horizontal polar response is in the form of a wide lobe but the vertical response is much narrower and usually angled downward. Units are often fitted with deflector plates to alter the response shape. Some are designed for long narrow areas thereby reducing penetration of side boundary walls, while others have split beams for covering two adjacent areas (Figures 102–104).

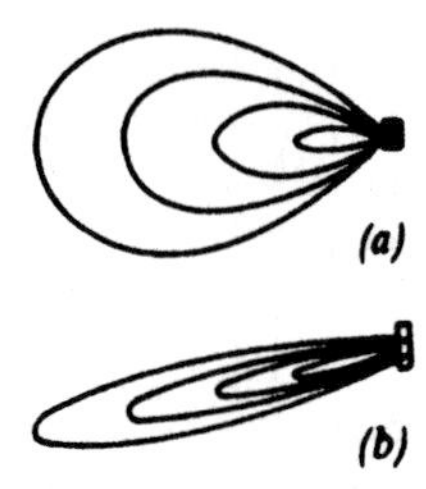

Figure 102 Polar diagram of microwave transmitter. Internal lobes indicate coverage when output is reduced. (a) Horizontal plane. (b) Vertical plane.

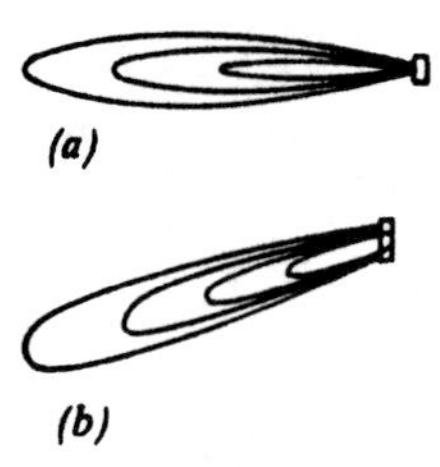

Figure 103 Narrow beam microwave for long narrow areas. This minimizes wall penetration and false alarms. (a) Horizontal plane. (b) Vertical plane.

Figure 104 Split beam microwave for covering two connected areas with one unit.

Microwave detectors can also be used for beam-breaking applications such as guarding perimeter fences (see Chapter 5).

Active infra-red beams

Infra-red radiation lies just below visible light in the spectrum. It is not a fixed frequency but actually has a wider frequency band than visible light, ranging from 10^{12} to 10^{14} Hz, or 1000 to 100,000 GHz, which is a wavelength of 300 μm to 3 μm. It can be generated by a variety of sources including a low-temperature blackened filament lamp, but the usual generator is gallium arsenide crystals. Beams are usually modulated with a low frequency such as 200 Hz to avoid defeating the system by directing an infra-red source at the receiver. If the modulation disappears, even if the beam is still being received, the receiver signals an alarm.

Unlike microwaves, infra-red radiation is affected by snow, rain and ice, and so must be operated well within the specified range if used outdoors. In addition, heaters are included with outdoor models to keep frost and condensation at bay. Maximum ranges vary from 40 ft (12 m) for the smaller units to 1000 ft (300 m).

The infra-red beam can be reflected from mirrors and directed through glass. While a reflected beam may seem an ideal way of protecting a perimeter fence, by using mirrors at each corner, the possibility of frost or condensation on the mirrors makes the practice questionable. Unlike the transmitter and receiver, the mirrors cannot be easily heated.

For indoor applications these problems do not apply, but losses are incurred at each reflection or glass penetration. These reduce the range as shown in Table 8. The figures show the range as a percentage of the original at each reflection, or penetration of 20 oz glass. More than three reflections reduce the range to 30 per cent and is not recommended.

Table 8

Number	*Reflections (%)*	*Penetrations (%)*
1	75	84
2	56	70
3	42	60
4	–	50

Passive infra-red detectors

The pyroelectric infra-red detector consists of a slice of doped lead zirconate titanate ceramic material having electrodes deposited on opposite faces, and which has two sensitive areas. A voltage is generated when any heat or infra-red radiation falling on these areas, varies in intensity. The device is arranged behind a lens or in front of a mirror that has facets so that a moving radiating object is focused first on one sensitive area and then on the other. This produces abrupt changes in received radiation which thereby produce the voltage.

Wavelengths to which the detector responds differs between different types of device. All cover the upper portion of the infra-red band, from about 20 μm upward, but some cut off around 6 μm, while others extend to 1 μm, which is in the radiant heat region.

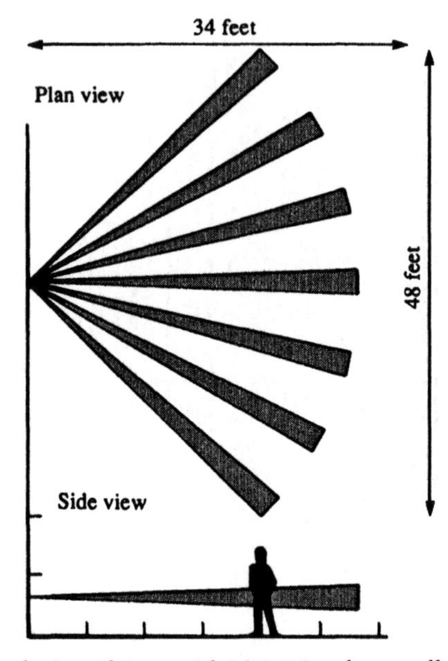

Figure 105 PIR sensor designed to avoid triggering by small animals.

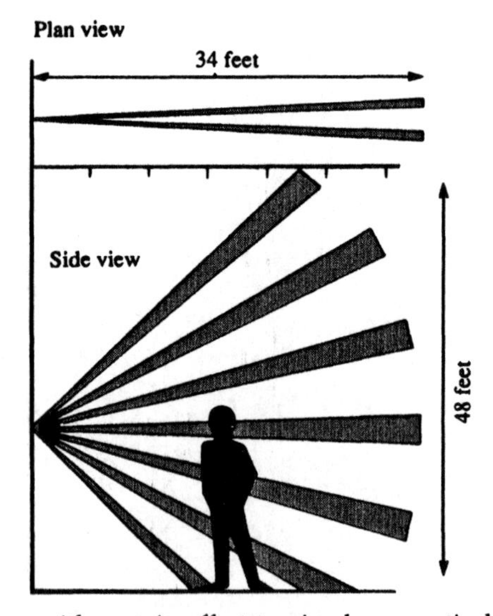

Figure 106 PIR sensor with curtain effect to give large vertical cover.

The detectors are capacitive and have a very high output impedance so a field-effect transistor (FET) is incorporated in the capsule; a source resistor is often included as well. Three terminal wires are provided for connection to the source, drain and common earth which is negative. Nominal supply voltage is 12 V.

Various ranges and polar diagrams are available depending on the configuration of the mirror or lens. Figures 105–112 show eight different diagrams for various applications.

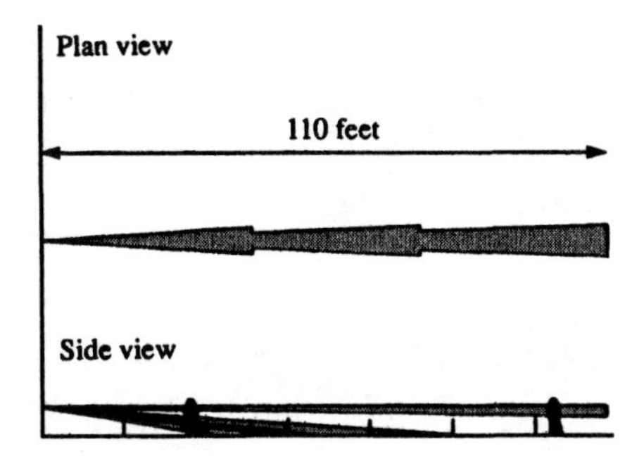

Figure 107 PIR with long narrow coverage for corridors and stairways.

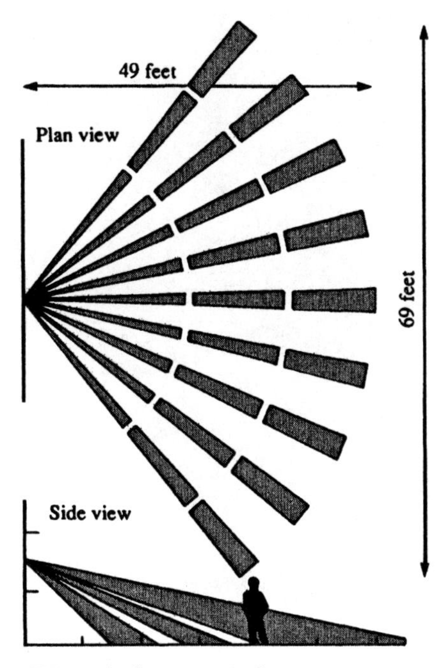

Figure 108 Large area PIR with three vertical zones.

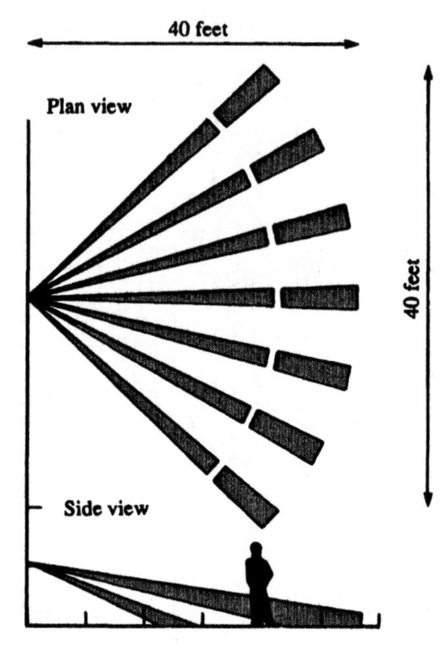

Figure 109 Medium range general purpose PIR.

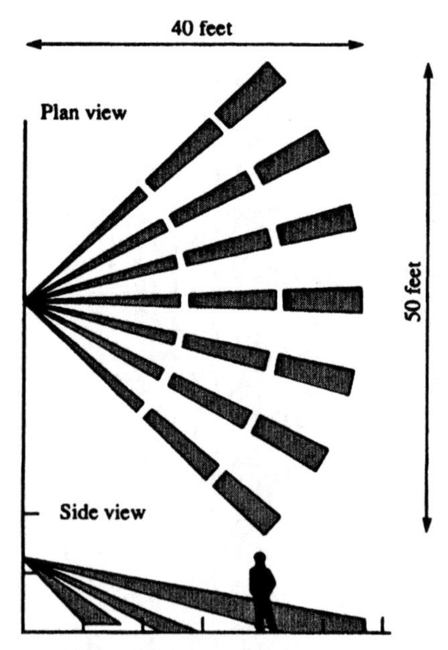

Figure 110 Medium range PIR with wide coverage area at a maximum range and a close zone for nearby detection.

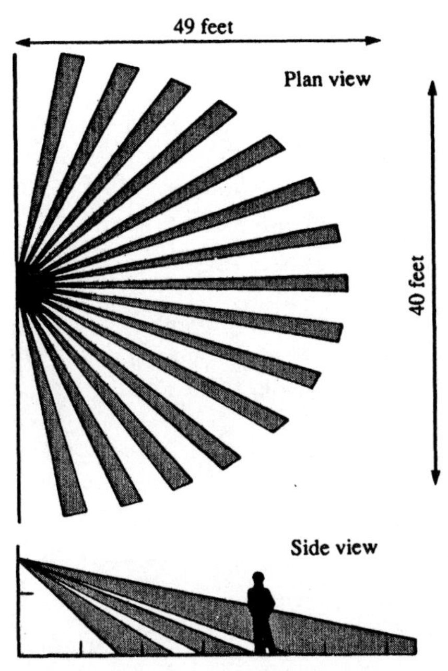

Figure 111 Wide-angle PIR for mounting on flat wall instead of usual corner position.

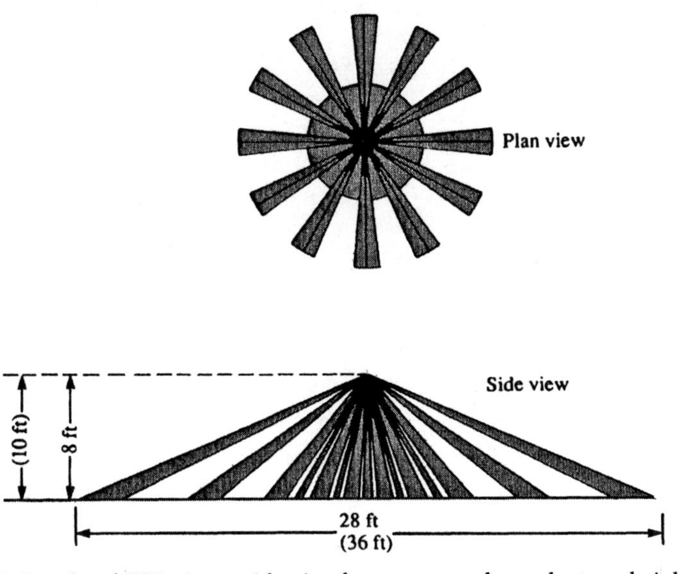

Figure 112 Overhead PIR gives wide circular coverage dependent on height.

Attenuation of sound with distance

Sound levels decrease with distance from the source generally at a rate of 6 dB for a doubling of the distance from a point source.

Table 9 shows the distances at which a given level at 3 ft (1 m) from the source decreases to 70 dB, 65 dB and 60 dB. Thus the distance over which a given sound volume will be maintained from a specified source can be found; alternatively, the required source level to obtain a given volume at a known distance can be determined.

A level of 60 dB is the minimum for fire alarms in a quiet environment. These figures can be modified by obstructions and absorbent surfaces and so are intended as a guide only. It is wise to plan for a higher volume than needed.

Table 9

dB at 1 m	*70 dB*		*65 dB*		*60 dB*	
	ft	*m*	*ft*	*m*	*ft*	*m*
83	14.6	4.4	26	8	47	14
86	21	6.3	37	11	67	20
90	33	10	60	18	107	32
93	47	14	84	25	150	45
96	67	20	120	36	214	64
100	107	32	186	56	333	100
103	150	45	266	80	480	144
106	210	64	380	114	668	203
110	340	102	600	181	1073	322
113	480	144	853	256	1520	456
116	680	204	1206	362	2150	645
120	1080	324	1916	575	3413	1024

Test equipment and tools

The type of test equipment needed by a professional installer depends on the types of installation and also whether faulty control panels are to be repaired or returned to the makers for service.

The first item and most important is a *multi-range test meter.* It need not have extended voltage and current ranges, but it should have a low-reading ohms range able to resolve small resistance variations in a loop having a poor connection. A high resistance range is also required to check for leaks.

Though not essential, a very useful instrument is a *direct-reading capacitance meter*. High values of capacitance are not important, but low ones in the tens of picofarads are. This can be used as described in an earlier chapter to check for cable and loop breaks. Briefly, the capacitance of a specific length of cable (say 1 m), of the same type as that having the break is measured. Capacitance of the defective cable is now measured, and the first reading divided into it. The result is the distance along the cable to the break.

A *microwave testmeter* can save many problems when installing and setting up this type of space protector. The signal level can be measured thereby ensuring the sensitivity is set to give adequate protection up to the perimeter but no leakage beyond. Some meters combine an infra-red range which with beam systems can check that sufficient energy is reaching the receiver with sufficient margin to ensure false-alarm-free operation.

Inertia sensor testers consist of a spring-loaded shaft with a nylon tip that recedes inside the handle when pressed against a structural member, until the pressure is relieved with a short sharp shock. The principle is similar to that of the spring-loaded centre-punch. The shock is calibrated and so can be used to check the response of inertia and vibration sensors and to adjust the sensitivity.

Programmers for digital communicators are available to enable the memory chip to be programmed with the central station number and the subscriber identification codes. The chips normally used are EPROMs (Erasable Programmable Read Only Memory), but NOVRAMs (Non-Volatile Random Access Memory), which can be reprogrammed at least 1000 times and will retain its memory with the power removed for over 25 years, are also being used. The programmer needs to be designed for a specific memory chip.

A *metal-detector* of the type that locates gas pipes and cables buried in the wall plaster can save problems when installing. It could also locate buried system wiring when fault tracing.

If repair work on circuit boards is to be undertaken a, *logic probe* and *logic pulser* may prove useful, as microprocessors and logic circuitry are being increasingly used in control units. The pulser can generate a single pulse or a train which can be injected into various parts of the circuit. The probe detects and displays logic 0 and logic 1 pulses and so traces data present at any point in the circuit including the i.c. pins.

A *sound-level meter* is invaluable for checking the sound levels of sounders at various distances. In particular, it is necessary to ensure that the required level is present at all points in fire system installations. An expensive model with many facilities, is not necessary, a simple direct reading dBA meter is all that is required.

The usual complement of hand tools should be included in the kit, including special screwdrivers that may be required for security screws. A medium-powered soldering iron should be chosen, as the miniature ones used for small electronic components may have insufficient heat capacity for soldering cables and larger workpieces.

A good *cable-stripper* that removes the insulation without damaging the wire is essential, do not rely on side-cutters or pliers. A *free-standing adjustable lamp* powered by rechargeable batteries is another must, as poor light is the enemy of well-made connections, and both hands need to be free.

An *electric drill* is also essential with a good selection of sharp drills, including a long masonry drill with an extension piece to get through thick outside walls. A 20 mm flat bit drill is useful for fitting ¾ in magnetic contacts as it is slightly oversize and so facilitates easy fitting.

For fire alarm installations a *hair-dryer* that achieves at least 135° F (57° C) is necessary to test heat detectors, and a good supply of sash cord or similar for generating smoke.

A *staple-gun* can be a time saver when wiring large installations, but there can be reservations. The wire staples can penetrate insulation, and it is not always easy to get the gun into confined spaces. If used it should be used with caution, and in conjunction with other types of fixing when they are more appropriate.

24 British Standards

The British Standard (BS) that applies to intruder alarm systems is the BS 4737; that which covers wire-free (radio) alarm systems is BS 6799; while that which has reference to fire alarm systems is BS 5839. All of these are lengthy documents having numerous modifications and up-dates that have been added over the years. The following is just a summary of the main points which concentrates on technical details, and is up-to-date at the time of going to press. Anyone such as an intending manufacturer, having a particular interest in a certain part, should consult the actual BS of which most large libraries have a copy.

Not all the recommendations have the support of those in the industry, and with most of the controversial ones the standard is considered to be not high enough. These though are in the minority, and the greater number are sound and should be adhered to. Insurance companies usually require any alarm system to be according to the relevant BS.

BS 4737

(Part 1)

Part 1 deals with the intruder alarm system, excluding the sensors.

Housing

The housing of the control unit, outside bells and other equipment should be of mild steel of not less than 1.2 mm gauge, stainless steel not less than 1 mm, or polycarbonate not less than 2 mm. All should have anti-tamper microswitches.

Wiring

When the system is set, an open-circuit or short-circuit of any sensor cable should trigger the alarm. It is also desirable, though not mandatory, that cables should be monitored when the system is not set.

The cables should be 0.5 mm^2 class 5, as described in BS 6360 (1981), with polyvinyl chloride (PVC) insulation not less than 0.25 mm, and sheathed with TM2 compound. Tinsel wire can be used in which case each strand should have a resistance of less than 270 ΩKm^{-1} at 20°C with PVC T12 insulation of not less than 0.3 mm. Total resistance of all circuits should be less than that which would reduce the voltage to below the required minimum at full load. Joints can be made by wrapping, crimping, soldering, clamping, by plug and socket, or wire-to-wire joint either insulated or in a junction box.

All wiring must be within the protected area or be mechanically protected where this is not possible. None should be run in the same conduit or trunking as mains wiring unless physically separated.

Control equipment

Control units should be within the protected premises and not be visible from the outside. All detection circuits should latch. Each zone should give either audible or visible indication of an alarm condition existing during setting or unsetting, or when testing the system.

It should not be possible to set the system in an alarm condition. Unsetting, should be by keys having no less than 200 differs (200 possible key profiles) so that the chance of a similar key being available is small. Alternatively another method having a similar security level can be used. (There is no mention at present in the BS of key-pads, although these would be covered by this alternative.) After an alarm, there should be a clear indication of which zone was triggered, and if more than one, which was the first.

Setting and exiting

Exiting should be by a timed delay circuit, or a setting control at or immediately outside the exit point. An audible warning should be heard over the whole exit route and immediately outside it. Shunt locks used for exiting should have at least 200 differs.

Unsetting

This can be by means of a lock-switch or door-switch, or any other means if within the protected area. Audible warning should be given during entry until the system is unset, but this is required only if a remote signalling device is connected to the system. A second delay is permitted to start after the first has expired, during which only a local sounder operates. If still unset after the second delay a full alarm should occur. If a sensor in any other part of the premises operates during the entry routine, a full alarm should sound. Circuits can be isolated by a shuntlock having more than 1000 differs, or by any other means of similar security level.

Alarm response

The system should respond to an alarm signal of not less than 200 ms but more than 800 ms. Response should be within 5 seconds. At least one warning device (sounder or remote signaller) should operate within 5 seconds.

If the premises are partly occupied when the system is set, and a sensor is thereafter triggered, an alarm should be signalled by a local sounder not more than 5 minutes later. If the system remains unset, a full alarm should sound not more than 5 seconds after this. It should not be possible for a system to be reset by a subscriber after an alarm, but only by the alarm company's engineer or an approved trained person.

Power

Power should be derived from the mains via an isolating transformer and correctly fused. It should not be supplied through a plug and socket.

Secondary standby batteries should have a capacity for not less than 8 hours operation, and should be recharged within 24 hours. Changeover in the event of mains failure should be automatic.

Primary batteries should have sufficient capacity to run for not less than 4 hours in the alarm condition and should have the date of installation marked on them. This applies also to secondary batteries that are not automatically recharged. System operating voltage should be 12 V, but higher voltages are permitted though not exceeding 50 V.

If the voltage becomes low, an alarm should be triggered. It should not be possible to set the system with low voltage.

Sounders

Housing should be as described earlier, in addition there should be no projections to which ropes or chains could be attached. At least two fundamental frequencies between 300 Hz and 3 kHz should be generated.

Output should be greater than 70 dBA at 3 m. This is measured by mounting the sounder on a solid support with at least 50 mm of the material surrounding it, and a counterweight at the rear. The assembly is suspended 1.2 m from the floor and a measurement taken on axis at 3 m distance. The reading should be greater than 65 dB in any direction. Absorbent material should be laid on the floor to prevent reinforcing reflections during the test.

Self-actuating bells should be used having a battery with a capacity of not less than 2 hours sounding. It should be rechargeable within 24 hours. A short-circuit across the charger should not discharge the battery. An alarm should sound if connections are changed or the tamper circuit is actuated.

The bell should be capable of functioning in an environment of –10° C to 55° C and up to 95 per cent humidity.

Internal sounders have the same requirements as external ones and should also be self actuated.

Remote signalling

Any British Telecom (BT) line used for remote signalling should be either a dedicated outgoing-calls-only line or ex-directory. It is recommended that it be buried underground or concealed over its whole length. It should be continuously monitored for faults.

For 999 diallers, the equipment should be triggered by any alarm signal longer than 200 ms, and should respond within 5 seconds. Once started, it should not be possible to interrupt the transmission.

If the sounders are delayed, the set delay will be reduced to less than 30 seconds if a telephone line fault is detected by the dialler. An alarm initiated by a panic button may not necessarily operate the sounders at all if it is deemed more prudent to rely on remote signalling to summon help.

Digital communicators should also be triggered by an alarm signal of not less than 200 ms, but the response should be within 1 second.

If contact is not established with the monitoring station within 1 min, the BT line should be released and the process restarted. If there is no contact after no fewer than three attempts, the sounders should operate in less than 10 seconds. If contact is made but no acknowledgement is received after not more than ten message transmissions, the process should be restarted.

Standby power supply batteries for a remote signalling apparatus should be of sufficient capacity to power it for at least five alarms if rechargeable, and should be recharged within 24 hours. Primary non-chargeable batteries should be able to power the equipment for at least twenty alarms. The date of installation should be marked on each battery.

(Part 2)

Part 2 refers to alarm equipment that is used only with deliberately operated sensors (panic buttons). Apart from exit routines, the conditions are virtually identical to Part 1.

(Part 3)

Part 3 covers the various sensors and their installation.

3/1 Taut wiring

This is used to protect walls and other structures that could be broken through. It should be of hard-drawn copper of 0.3–0.4 mm gauge. Polyvinyl chloride (PVC) insulation should be 0.2–0.3 mm thick.

Fixing points should be no more than 600 mm apart, and the spacing between adjacent runs no more than 100 mm. If run in tubes or grooved rods the spacing to adjacent ones should be the same, and they should be supported at not more than 1 m intervals. If recessed into supports, the amount of recess should be more than 5 mm but less than 10 mm from the end of the tube.

The wire emanating from the tube ends should be supported at no more than 50 mm from the tubes. Less than 50 mm of displacement should be necessary to break the wire and sound the alarm.

3/2 Foil on glass

Foil should be less than 0.04 mm thick and less than 12.5 mm wide. If a single-pole configuration is used, the foil can be laid as a rectangle between 50 mm and 100 mm from the edge of the glass. It can be run as a single strip not less than 300 mm long if the short window dimension is less than 600 mm. In this case it would be run through the centre, parallel to the long edges. It can also be laid as parallel strips not more than 200 mm apart, the ends terminating 50–100 mm from the edge. Unframed glass can be protected with a loop not less than 200 mm by 200 mm.

3/4 Microwave sensors

The area covered by a microwave sensor using the Doppler effect is defined as that in which an alarm is triggered by a person weighing 40–80 kg moving at between 0.3 and 0.6 m/s through a distance of 2 m, or 20 per cent of his radial separation from the sensor. The sensor should respond only to signals of over 200 ms.

3/5 Ultrasonic sensors

The area covered by an ultrasonic sensor using the Doppler effect is that in which an alarm is triggered by a 40–80 kg person moving through a distance of 2 m at any speed. Response should be to signals over 200 ms.

The frequency must be higher than 22 kHz. The sensor should be capable of operating at temperatures of 0°C – 40°C and at humidities of between 10 per cent and 90 per cent. A warning is given that nuisance or health hazard is possible from intense ultrasonic radiation. Safe limits are still under consideration, but it is recommended that the device does not radiate when persons are lawfully nearby.

3/6 Acoustic detectors

The sensitivity control should be set to trigger when a sound is greater than 15 dB over the ambient noise, or at a maximum of 85 dB. There should be no more than 6 dB variation within those limits. The detector should respond to signals longer than 5 seconds in any 30 second period, or any input greater than 120 dB for 100 ms.

3/7 Passive infra-red detectors (PIRs)

The area covered is defined as where an alarm is triggered by a 40–80 kg person moving laterally through a distance of 2 m at any speed. It should not generate an alarm signal when a target that is equivalent to a 40–80 kg person fills the view of the detector, and is heated at an even rate of less than 0.1°C per second.

When a remote signalling device is included in the system, it should not be connected until there has been at least 7 days trouble-free operation of the PIR, with the exception that the detector is replacing one of similar type or there is an urgent need for immediate maximum security.

3/8 Capacitive detectors

These are sensors that operate when there is a change of capacitance or a rate of change in the proximity of the device. Their area of detection is defined in the same way as for a microwave sensor namely, as that in which an alarm is triggered by a person weighing 40–80 kg moving at between 0.3 and 0.6 m/s through a distance of 2 m.

3/9 Pressure mats

The pressure required to operate a pressure mat is that applied by a disc of 60 mm in diameter ±5 mm, pressing at right-angles to the mat with a force greater than 100 N. It should not be actuated by a force applied by the same disc, of less than 20 N. (1 Newton the force applied to 1 kg to accelerate it 1 metre per second per second. It is also equal to 100,000 dynes.)

3/12 Beam interrupters

An alarm should not be generated if a beam is reduced by less than 50 per cent. The alarm signal should be greater than 800 ms in length for an interruption longer than 40 ms, but should not be triggered for one less than 20 ms. In all cases the source should be modulated to prevent the receiver being affected by another source accidental or deliberate. (The type and frequency of the modulation is not specified.)

3/14 Deliberately operated sensors (panic buttons)

Three types are described:

1 that requiring a single force on one element;
2 that needing two simultaneous forces on two different elements; and
3 those operated by two consecutive forces applied to different elements, the first being maintained while the second is applied.

The sensors can be latching or non-latching.

For a manually operated device, the force required should be within 4–5 N as applied by a 6 mm diameter disc. For pedal operated devices the force should be within 5–8 N applied by a 12 mm diameter disc.

3/30 Cables

Cables for intruder systems should be either plain or tinned annealed solid copper not less than 0.2 mm^2, with a maximum resistance of 95 Ωkm^{-1}. A *pro rata* resistance should apply for cables up to 0.5 mm^2.

Stranded cables should not be less than 0.22 mm^2. Any strand joints should be brazed or hard soldered and not be less than 300 mm apart. Tensile strength of these should not be less than 90 per cent of adjacent continuous cable.

Cores should be no fewer than 7 strands of 0.2 mm conductor. Insulation should overlay the wire but not adhere to it and so prevent a clean strip. Sheaths should likewise not adhere to cores. Cable sheaths should not be less than 0.4 mm thick. Insulation resistance should be greater than 50 MΩ for 1 kM at 20° C.

(Part 4)

4/2

This deals with the security of the installing, company, and maintenance. Staff must be thoroughly vetted before being taken on to deal with client's security systems, and must receive adequate training to ensure competence. The confidentiality of all records and details must be totally preserved.

Maintenance visits and tests should be made not more than 12 months apart for mains-operated systems using local sounders. For battery-powered systems, the maximum time between visits should be 6 months. When remote signalling equipment is included, the visits should also not be more than 6 months apart.

On each visit the following items should be checked: The record book for problems arising since the last visit; a visual check of the system and any changes in layout or stock storage that could affect it; all sensors tested;

flexible cables examined; power supplies including standby batteries; control equipment; sounder; remote communicators and a full operational test.

Accurate records of all alarm events, maintenance and repairs should be kept.

BS 6799

This Standard describes wire-free (radio) alarm systems. There is no mention of the frequencies, or radiated powers to be used, as these are regulated by the Department of Trade and Industry. At present the DTI has allocated 173.225 MHz for use by intruder alarms providing the transmitter is crystal controlled at 25 kHz channel spacing. No license is required for an effective radiated power (ERP) of up to 1 mW. For longer ranges or in electrically noisy environments, up to 10 mW is permitted, but this requires a license. Another suitable band for longer range is 458.5–458.8 MHz in which up to 500 mW ERP is permitted without a license. These bands are for transmission of data and not for speech.

The BS 6799 describes five levels of security in systems using wire-free sensors. The need for such a choice arises from the power supply limitations. Sensors need to be reasonably small, yet each must contain a battery having a life of some 6 months. Continuous transmission to provide 24 hour monitoring is thus not feasible. The five different levels reflect different compromises between monitoring and battery life.

1 The sensors transmit only when actually triggered and also signal when the battery is getting low, having capacity for only 7 further days of operation.
2 As 1, but sensors also transmit a code identifying which one has been triggered. Thus it indicates which part of the premises is affected.
3 As 2, but the receiver monitors the channel for interfering or blocking signals lasting for more than 30 seconds, and gives a warning of such.
4 As 3, but the sensors transmit a return-to-normal signal after an alarm. They also regularly report by transmitting their status and battery condition at no longer than 8.4 hour intervals. If no status report is received in an 8.4 hour period the receiver signals a fault condition, if there are no reports for three consecutive periods a full alarm is sounded. A fault indication is given if a low-battery signal is received in two consecutive reports.
5 As 4, but the reports are transmitted at 1.2 hour intervals.

The Standard also outlines the advantages and disadvantages of radio sensors. Advantages are the obvious ones of eliminating wiring, no damage to historic decor, portability and easy addition or changes to the system. Disadvantages are the possibility of blocking or interference, limited monitoring and no anti-tamper protection, as well as frequent battery replacements for each sensor.

BS 5839

(Part 1)

This Standard covers fire alarm systems, and now includes information contained in the BS 3116 (1974) which dealt with automatic fire alarm systems and was previously separate from BS 5839.

Self-contained smoke alarms are recommended only for domestic single-family dwellings even if several are interlinked. In multi-occupied buildings, whether business or domestic, alarm systems should be coordinated and if possible interlinked.

Detectors

While these must be removable for cleaning and service, special tools should be required to do so wherever there is a possibility of vandalism and malicious removal.

It has been found that escape routes are rendered ineffective if visibility due to smoke is less than 10 m, because people are unwilling to walk through greater densities. So, smoke in those locations should be detected as early as possible. Smoke detectors should therefore be positioned on all escape routes and stairway ceilings, to give the earliest possible warning.

Wherever radioactive materials are used at work, regulations require notification to the Health Authority. However, this is only the case if the activity is greater than 4 MBq, and the dose rate is more than 1 μSv.h^{-1} at 100 mm. Most smoke detectors use Americium 241 and are well below this level so no notification is required.

The recommended mounting positions are as those described in Chapter 22.

Zones

These should be separated from each other by a natural fire barrier and should not exceed 2000 m^2. Each storey should be a separate zone as also should stairwells, liftwells and other flue-like channels. The search distance, that is the distance someone would have to search before having visual indications of a fire should be less than 30 m.

Any building having less than 300 m^2 of floor area can be regarded as a single zone even though it may have more than one storey.

Circuits and cables

Circuits should be monitored so that a fault is registered in less than 100 s. Two simultaneous faults should not remove protection from an area greater than 10,000 m^2.

If a ring circuit is used, any open-circuit should be indicated as a fault in less than 60 min. Although the circuit is still fully operative, there is now no back-up and a second open-circuit would disconnect a number of sensors between the two breaks.

A short-circuit should be indicated as a fault within 100 s, but if there are isolators in the ring to limit the effect of any such short-circuit, the indication and remedial action can be within 60 min.

Wherever operation is required for as long as possible after a fire has broken out, such as for sounder wiring, the cables should be of the heat-resistant type or otherwise protected against heat. Cables should be larger than 1 mm^2 for solid wire, or larger than 0.5 mm^2 for stranded. Those less than 1 mm^2 should not be drawn into conduit unless it is twin twisted in which case down to 0.5 mm^2 can be so used.

Fire alarm cables should be segregated from all those for other purposes. This should be done by running in conduit, ducting, trunking, or if exposed, by a greater than 300 mm separation from other wiring, or by using special cable for unsegregated wiring. If none of these is possible, the wiring should be labelled at intervals of less than 2 m.

All joints other than at sensors, sounder and control units should be made in junction boxes and labelled *'FIRE ALARM'*.

Manual call points

There should be a maximum of 3 second delay in sounding the alarm after a manual call point has been operated. The maximum distance from anywhere in the building to a call point should be 30 m, but this should be less if inflammable materials or processes are being used.

Call points should be clearly visible and so should be of contrasting colour to the background and clearly labelled. They should not be fully recessed into a wall if approach may be made from the side, as in a corridor. Partial recessing is permissible providing the visible side profile is not less than 750 mm^2.

Sounders

The levels should be higher than 65 dBA at any point in the premises, or at least 5 dBA above any ambient noise lasting for more than 30 seconds if this is greater. For premises housing sleeping persons, the level should be at least 75 dBA at the bedhead with room doors shut. Sounders should be sufficient to maintain these levels if some are inactivated by fire. There should be at least one in each fire compartment, and a minimum of two in any installation however small.

Frequencies generated by sounders should be between 500 Hz and 1 kHz, but can be outside that band if there are ambient noise frequencies present that could mask the alarm sounder. All sounders should be the same unless there is some specific reason why any should differ. The tone should be quite distinctive and unlike any other signal such as one used to signal tea-breaks or an end-of-shift.

There should be no coded signals, such as one ring for the first floor and two for the second, as these could be misheard or misunderstood. Two-stage alarm soundings are permissible, that is a first sounding in the immediate vicinity of the fire and above it, and a second one shortly afterward in other areas if it is evident that the fire is not under control and is likely to spread.

An alarm restricted to staff areas is also permitted where panic may ensue if a public alarm was sounded. The condition is that the number of staff be adequate to give verbal warning and supervise an orderly evacuation. A nursing home could be an example.

Sounders should not be so placed as to affect telephone communication. Details being given to the fire brigade have sometimes been obscured, and even some tones in the alarm sound have confused the BT dialling tone recognition system.

Power

Power should be derived from the mains via an isolating switch fuse which is coloured red and prominently labelled *'Fire Alarm Do Not Switch Off'*.

Standby batteries should be of the rechargeable type with a minimum life of 4 years. They should have sufficient capacity to operate the system for more than 24 hours, and the charger should be capable of recharging them in less than 24 hours. Car batteries should not be used.

(Parts 2–5)

These give details of manufacturing standards for equipment and so is of interest mainly to manufacturers. The parts are: 2, Manual call points; 3, Auto-release systems; 4, Control equipment; 5, Optical smoke detectors.

Conclusion

It is evident that two different teams put together BS 4737 and BS 5839 as there are some notable differences. Although they serve quite different situations, intruder and fire alarms have much in common. The Standards each have good points that could be applied beneficially to the other.

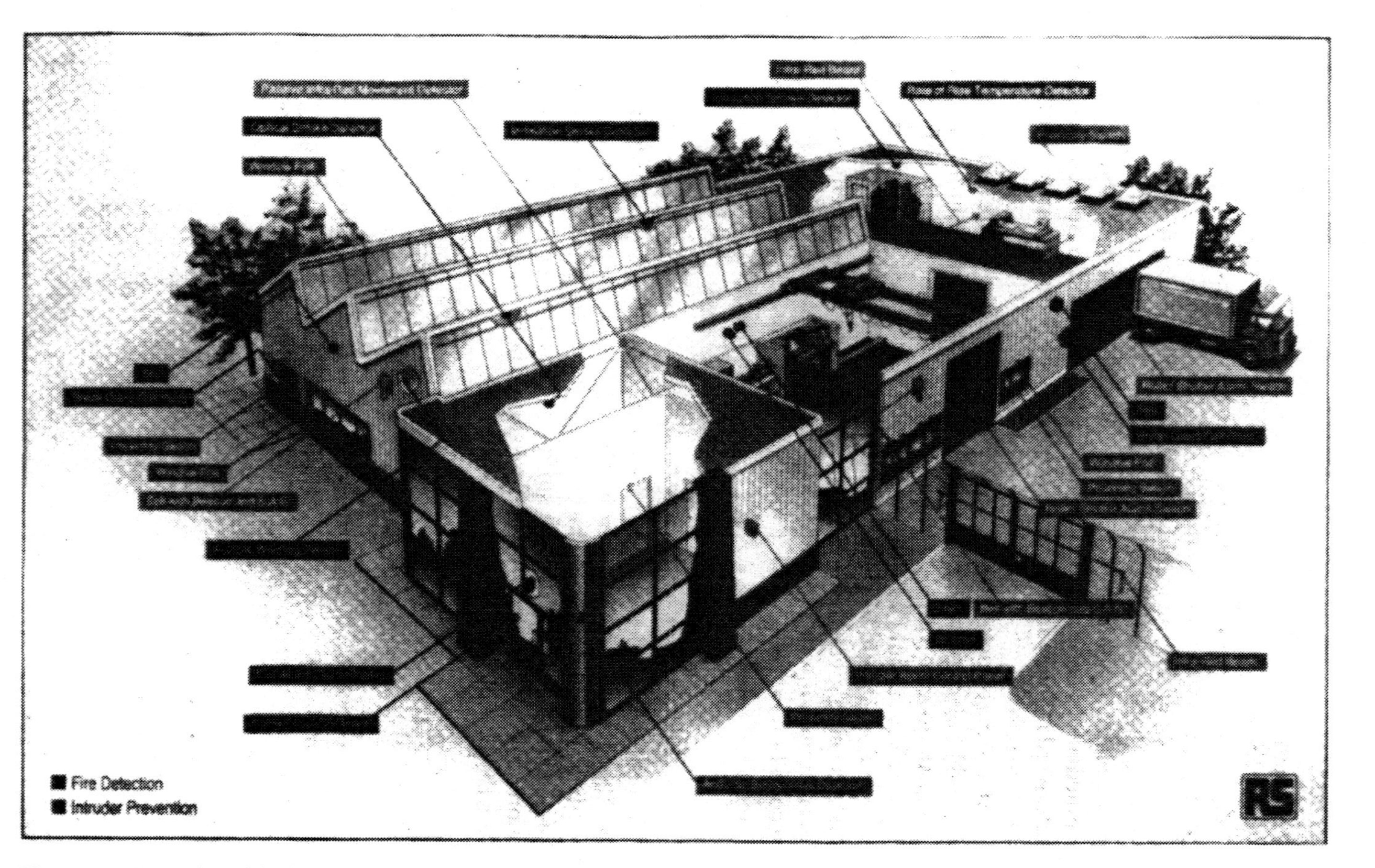

Figure 113 Reproduced by kind permission of RS Components.

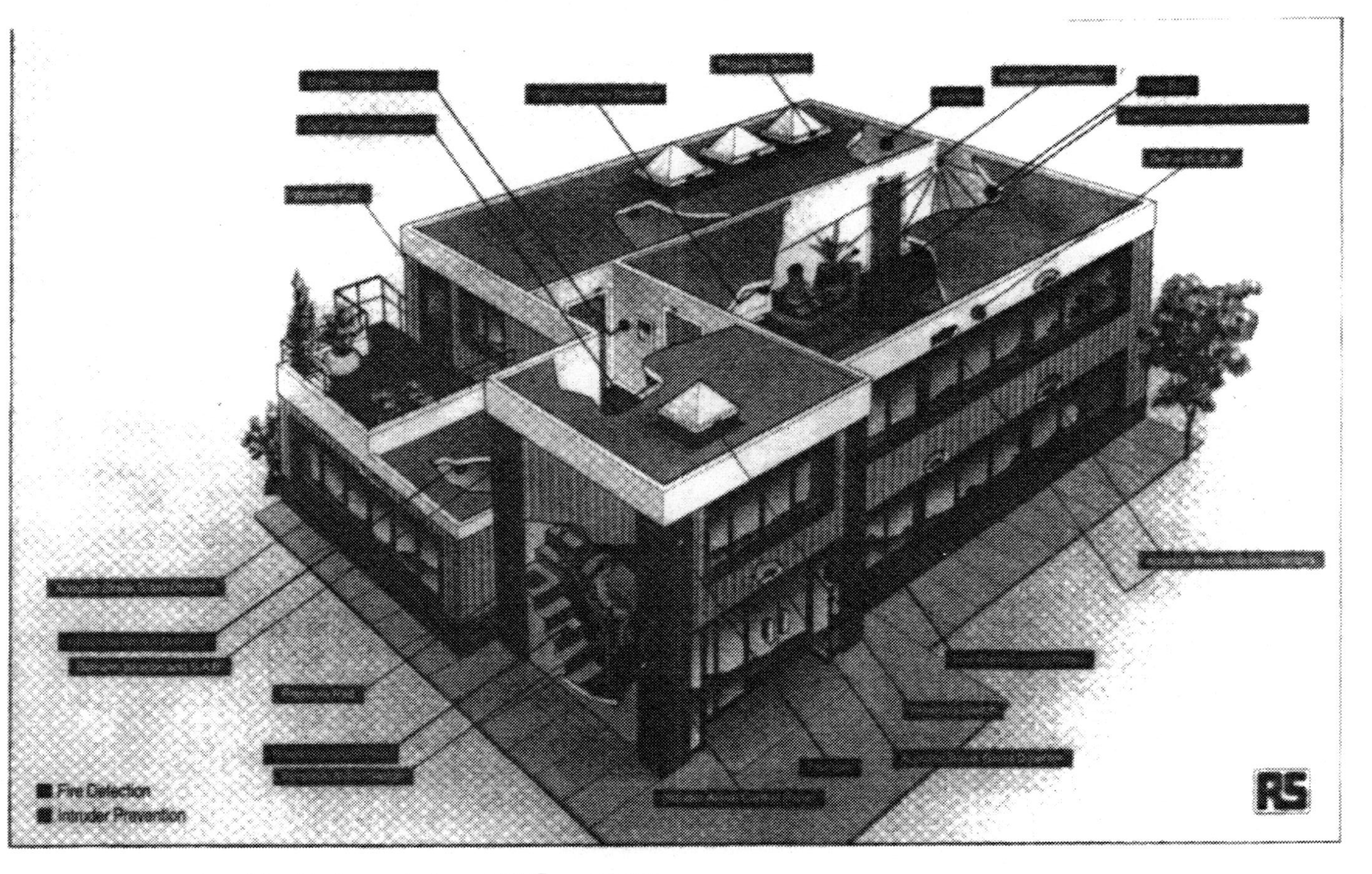

Figure 114 Reproduced by kind permission of RS Components.

Index

Printed in the United Kingdom
by Lightning Source UK Ltd.
111292UKS00001B/89

9 780750 642361